Cambridge Monographs in African Archaeology
90
Series Editors: Laurence Smith, Brian Stewart and Stephanie Wynne-Jones

AF687232

Swahili Urbanisation, Trade and Food Production

Botanical perspectives from Pemba Island, Tanzania, AD 600–1500

Sarah Walshaw

BAR International Series 2755
2015

Published in 2016 by
BAR Publishing, Oxford

BAR International Series 2755

Cambridge Monographs in African Archaeology 90
Series Editors: Laurence Smith, Brian Stewart and Stephanie Wynne-Jones

Swahili Urbanisation, Trade and Food Production

ISBN 978 1 4073 1416 7

© S Walshaw and the Publisher 2015

The author's moral rights under the 1988 UK Copyright, Designs and Patents Act are hereby expressly asserted.

All rights reserved. No part of this work may be copied, reproduced, stored, sold, distributed, scanned, saved in any form of digital format or transmitted in any form digitally, without the written permission of the Publisher.

BAR Publishing is the trading name of British Archaeological Reports (Oxford) Ltd. British Archaeological Reports was first incorporated in 1974 to publish the BAR Series, International and British. In 1992 Hadrian Books Ltd became part of the BAR group. This volume was originally published by Archaeopress in conjunction with British Archaeological Reports (Oxford) Ltd / Hadrian Books Ltd, the Series principal publisher, in 2015. This present volume is published by BAR Publishing, 2016.

Printed in England

BAR
PUBLISHING

BAR titles are available from:

BAR Publishing
122 Banbury Rd, Oxford, OX2 7BP, UK
EMAIL info@barpublishing.com
PHONE +44 (0)1865 310431
FAX +44 (0)1865 316916
www.barpublishing.com

Abstract

In this monograph I examine Swahili plant subsistence and food production patterns through the analysis of macrobotanical remains from four archaeological sites on Pemba Island, Tanzania, dating to A.D. 700-1600. Specifically, I compare towns and villages before and during the emergence of stonetowns, settlements characterized by stone/coral household and ritual architecture, which have been described as urban based on their roles as economic, political, and religious centers along the eastern African coast. Swahili stonetowns are hypothesized to have exerted political control over the immediate hinterland for the purposes of obtaining trade items and staple goods, including plant products. Based on ethnohistoric reports, a wide variety of collected and cultivated plants has been proposed as being central to Swahili consumption and production economies, including trees in mangrove habitats, coconut, sorghum, pearl millet, and Asian rice. Moreover, it is assumed that stonetowns obtained plant products, including staple grains, from the countryside and were not themselves primary food producers. I test these assumptions directly against the archaeological record in this first comprehensive study of ancient Swahili plant foods.

The rich archaeobotanical record on Pemba contained wood, parenchyma, coconut, fruits/nuts, cotton, pearl millet, finger millet, sorghum, Asian rice, pea, and mung bean, among others. Archaeobotanical comparisons of contemporaneous towns and villages suggest relative economic independence. Before A.D. 1000, this independence is inferred through taxonomic representation. The smaller village site showed reliance upon more local fruits and starchy plants, while the contemporaneous town residents had access to grains (African millets and Asian rice), and more coconut and cotton. After A.D. 1100, stonetown and village plant assemblages resemble one another taxonomically, relying on Asian rice, coconut and other fruit/nut taxa, and cotton, yet they exhibit economic independence in terms of production. An analysis of crop processing strategies reveals a predominance of household-based production in both the village and stonetown. Thus, it appears that the stonetown was not being provisioned by rural farming settlements. Preliminary animal food and craft production data from these same sites lend further support for household production and the relative economic autonomy of Swahili settlements in northern Pemba Island.

Acknowledgements

This work began as my dissertation project, and I am grateful to my committee members at Washington University in St. Louis for offering excellent advice on a topic that was, in most respects, worlds apart from their own research. First, I thank Gayle Fritz for over six years of mentorship as my supervisor and for supporting my pursuit of an African dissertation topic. Fiona Marshall has given me outstanding advice, introduced me to Adria LaViolette, and helped me get acquainted with Swahili studies. I greatly appreciate the comments of and discussions with the remainder of my committee, Dave Browman, T.R. Kidder, Kenneth Olsen, and Susan Rotroff, who brought their diverse expertise to this study. I am grateful to my post-doctoral mentor Cathy D'Andrea at Simon Fraser University for supporting the ethnoarchaeological research that informs my interpretations here, and for providing continued research support and guidance. My colleagues in the departments of Archaeology and History at SFU have been highly supportive and have generously provided valuable feedback on my work. The last phase of writing took place in Oxford, where I benefitted enormously from interactions with colleagues and fellow visitors at the African Studies Centre and the Archaeobotanical Research Group at the Institute of Archaeology, University of Oxford.

My heartfelt thanks and appreciation extend to the directors and participants of the Pemba Archaeological Project. Director Adria LaViolette has been remarkably supportive of my work throughout all its stages, and has been patient with my academic learning curve. Jeff Fleisher generously helped with many aspects of the fieldwork associated with this project, and was a master problem-solver on Pemba, as well as a great source of project information ever since. They provided a rich work environment in the field and I have continued to benefit from their wisdom and friendship. I thank Hamad Omar and the Tanzanian Division of Antiquities for approving and supporting the Pemba Archaeological Project and my archaeobotanical work. I am sincerely grateful to Hajj Mohammed Hajj, who extended his expertise in the coastal environment and culture as well as the gift of friendship. This study benefited from the support of numerous Tanzanian faculty (particularly Bertram Mapunda, of the University of Dar es Salaam) and students, project specialists (especially Chet Cain), local archaeologists, project graduate students, and Tumbe residents. I am especially indebted to all who assisted me in flotation, botanical collection, and ethnographic interviews, who are too numerous to name here. I am also grateful to the five Pemban farmers and one traditional healer who shared their expertise during the course of this study.

I am honored to have worked with several botanical experts during this project. Roy Gereau of the Missouri Botanical Gardens has greatly expanded my knowledge of African plants, and helped me to identify my botanical comparative material. Deborah Lalumondier allowed me access to palms fruits from the MBG's famous Climatron. I am very grateful to Dorian Fuller (University College London) and Cathy D'Andrea for their contributions to my archaeobotanical education, without which this study would not have been successful. This work benefited from the sage comments provided by members of the Archaeobotanical Discussion Group at the University of Oxford, particularly Amy Bogaard. Any errors or inadequacies that remain are solely my responsibility.

This work was funded by the National Science Foundation, who provided a grant to LaViolette (Standard Grant #BCS0138319) and a dissertation support grant to Fritz and Walshaw (#0431137). I am grateful to the Natural Science and Engineering Research Council of Canada for a graduate student fellowship, and to Washington University in St. Louis for financial support and excellent resources for graduate students. I thank my brilliant colleagues in the PhD program in Anthropology for rousing debates and unflagging support, especially: Maria Bruno, Lisa Hildebrand, Liz Horton, Kate Grillo, Karla Hansen-Speer, Dawn Kaufmann, Kristin Arntzen, Monte Abbott and Kevin Hanselka.

I thank Stephanie Wynne-Jones for her insightful and helpful comments on my original dissertation and for proposing that it could be a good candidate for a BAR report. The editor, David Davison, and staff at BAR provided key assistance with good cheer.

Finally, I owe a huge amount to my family. My parents, Richard and Donna Walshaw, have always encouraged my scholarly pursuits. My brothers, Tim and Greg, have long suffered my nerdiness and travel rants, often with only a T-shirt from a foreign land as compensation. My sisters-in-law Jill and Katie have given their love and understanding, as well as their wisdom. Most of all, I am truly blessed with the encouragement and understanding of my husband, Jim Johnson, and our bright and sweet daughter Danica, who sorts shells like a future scientist.

Table of Contents

List of Tables

List of Figures

CHAPTER 1: INTRODUCTION

The biological need to acquire, prepare, and consume foods is universal, but among human societies foods have social as well as nutritional functions and implications (e.g., Goody 1982; Palmer and van der Veen 2002). In ancient societies, the adoption of agriculture represented a key change in lifestyle, and the resulting potential for food surplus has been credited with the emergence of social inequality and the development of complex societies (Childe 1951; Braidwood 1981; Smith 1995; Trigger 2003). The origin of cities has typically been associated with a specialized food production system managed by a central authority (Galvin 1987; Zeder 1991, 1999, 2003), in which villagers provided urban administrators with food given as tax or tribute (Hastorf 1993). The character of food production in non-archetypal urban systems, however, remains poorly understood.

Recent archaeological scholarship has come to identify different urban forms and trajectories around the world (Sabloff 2005), particularly in Africa (McIntosh and McIntosh 1993; LaViolette 2005), where some early cities did not follow the classic model, and where recognizable city forms were attributed to external, not indigenous African, efforts. African cities have recently been recognized as exhibiting novel characters, lacking some of the traditional attributes of classical cities but functioning as the economic, political, and/or religious heart of a region. Moreover, African food production systems followed different trajectories than those documented elsewhere (Marshall 1998; Marshall and Hildebrand 2002; Neumann 2003). Therefore, models of centralized food production developed for hierarchical systems may not apply to those African cities in which rank was based on age or kinship, and where African domesticates likely dominated. Investigations of Africa's unique pathways to food production and urbanism have the potential to reveal novel examples of urban food production. Considering the impact that food production had on local subsistence and trade economies, as well as political and social relations in and among communities, archaeologists and historians require an updated understanding of how food production practices changed to support urbanization.

The Swahili represent one such example of Sub-Saharan African urbanism. Swahili stonetowns bearing stone and coral houses, mosques, and occasionally city walls appear along the eastern African coast by A.D. 1000, probably functioning as marketplaces and Islamic centers as connections intensified between Africa and the Indian Ocean world. The nature of Swahili political and economic life has recently received much attention from archaeologists working in the region (Horton and Middleton 2000; Kusimba 1999; LaViolette and Fleisher 2005). Historically, Swahili stonetowns were considered to be non-African, created and controlled by Arab centers (Hull 1976:120; Kusimba 1999:51). Recent investigations, however, have revealed the distinctly African foundations of Swahili language (Nurse & Spear 1985), society (Allen 1993), and material culture and architecture (Horton 1996). Thus, Swahili patterns of urbanization must now be reconsidered in the context of local, indigenous precursors to urbanism (Fleisher 2001). Moreover, models of Swahili urbanization must begin to incorporate the lifeways of those living in rural villages and the less wealthy quarters of stonetowns, people who have been understudied and undertheorized in Swahili archaeology (Fleisher and LaViolette 1999a; LaViolette and Fleisher 2005). Agricultural production lies at the intersection of rural and urban economies, and as such can offer a unique perspective on Swahili urbanization.

Investigating agricultural systems is one of the key contributions of archaeobotany, the study of plant remains from archaeological sites. The examination of how plant food resources were managed, distributed, and used informs archaeologists about many different aspects of past societies, including which plants were cultivated as opposed to collected from the wild, how surplus food supplies supported the rise of complex societies, and who could obtain wealth and status through the accumulation and trade of plant products. But, like urbanism, current models of early agriculture are largely limited to plants important in Southwest Asia and Europe (wheat, barley), and therefore are not applicable to considerations of early food production in many parts of the globe. New studies of crops vital to regions outside of Eurasia are crucial to the successful application of archaeobotany to archaeological investigations worldwide, particularly in Africa.

In this study I reconstruct plant subsistence patterns at four Swahili archaeological sites on Pemba Island, Tanzania (seventh to sixteenth centuries A.D.), with the ultimate goal of characterizing how food production practices changed during urbanization after the eleventh century. I use the example of two Swahili towns (seventh to tenth century Tumbe, eleventh to sixteenth century Chwaka) and their neighboring rural villages (eighth to tenth century Kimimba, fourteenth to sixteenth century Kaliwa), on Pemba Island, Tanzania. For each time period, I compare and contrast the relations between the town, defined as the larger (> 10 ha), more complex and wealthy settlement, and the village (<2 ha), characterized by a smaller, less dense habitation with less evidence of wealth. While the earlier town of Tumbe is large and demonstrates a connection with the Indian Ocean world, this settlement was distinctly different from the later stonetown of Chwaka, which was more densely settled and likely had a larger impact on the political and geographic landscape (LaViolette et al. 2003, 2004). Thus, the contrast I draw is not between two comparable urban entities and their hinterlands, but rather between the pre-urban (Tumbe and Kimimba) food production system on Pemba and that occurring during incipent urbanization (Chwaka, Kaliwa).

I conducted archaeobotanical recovery in conjunction with the Pemba Archaeological Project, led by Adria LaViolette (University of Virginia), Jeff Fleisher (Lehigh University) and Bertram Mapunda (University of Dar es Salaam). This ongoing project was designed to discover the nature of the lives of the "Swahili majority", those who lived in relatively impoverished earth-and-thatch households located in towns and villages, and to determine how economically integrated the towns and countryside may have been (LaViolette et al. 2003). The archaeobotanical data presented here are part of a larger corpus of local production and consumption data being generated by this multidisciplinary project (Cain 2005; LaViolette and Fleisher 2005; LaViolette et al. 2003, 2004).

In this study, I use ancient plant remains recovered from archaeological deposits to reconstruct past subsistence and plant food production activities. While identifiable plant remains are considered rare at tropical sites in general (Hather 1994; Pearsall 2000), and in eastern Africa in particular (Wetterstrom 1991; Young and Thompson 1999), the Pemban archaeobotanical record was rich in charred wood, seeds, parenchyma, and fragments of fruit stones and nutshell. This excellent preservation of plant remains is most likely due to several factors, including the sweeping of cooking and household debris into fires, the accumulation of materials in middens, and the rapid burial of middens in a densely populated environment, particularly at Chwaka (Walshaw 2005). The Swahili cultivated grains that have distinctive morphologies observable under low magnification, such as Asian rice and the African grains pearl millet, sorghum, and finger millet. I also report the presence of fragments of starchy fruit and tuber plants commonly associated with African diets, which were more problematic to identify to taxon. The Pemban archaeobotanical record reported here represents one of the most comprehensive data sets currently available for sub-Saharan Africa, and as such presents an excellent opportunity to study changes in local plant food economies during Swahili urbanization.

This book is divided into nine chapters. The following chapter (Chapter 2) models urban agriculture and reviews the archaeological literature on urban food production in sub-Saharan Africa. This chapter ends by proposing a model of expected urban food production on Pemba, drawn from a consideration of how Eurasian models might be adapted to fit African plants and societies. Chapter 3 provides an overview of the geographic and cultural landscape of the Swahili coast, focusing on trends observed in the study area of Northern Pemba using data from recently published (Fleisher 2003; LaViolette and Fleisher 2005) and unpublished literature (LaViolette et al. 2003, 2004). At the close of Chapter 3, I adapt the model of urban food production put forward in Chapter 2 to the cultural and botanical character of Swahili food production. I report my methods of data collection in Chapter 4, including ethnographic and botanical field collection methods and the collection, processing, and analysis of archaeobotanical samples conducted both in the field and laboratory. Archaeobotanical data presentation is spread over two chapters (5 and 6) that group sites according to chronology. Data from the first millennium town of Tumbe and the nearby village of Kimimba are considered together in Chapter 5, while the second millennium data from the stonetown of Chwaka is presented in Chapter 6 alongside the village data from Kaliwa and the eighteenth to nineteenth century upper occupation sequence at Tumbe. In Chapter 7, I systematically compare the Pemban archaeobotanical data against the model of urban food production laid out in Chapters 2 and 3. In Chapter 8, I discuss how the local food production patterns described in Chapter 7 intersect with data from other elements of the local and trade economies on Pemba, including pastoralism and the production of crafts such as pottery, iron, and cloth. Finally, I summarize the major findings of this study in Chapter 9, and present an interpretation of how Swahili food production influenced, and was influenced by, urbanization on northern Pemba Island.

CHAPTER 2: URBANIZATION AND THE RISE OF SPECIALIZED FOOD PRODUCTION

In this chapter I focus on how the urbanization process affects food production and how this is manifested in the archaeological record. I discuss briefly how archaeologists have approached studies of urbanism and urbanization. I then move to a survey of models of food production in urban contexts, and examine recent reinterpretations of classic models of crop processing to adapt them to the urban question. I provide examples of how food production may have functioned in other sub-Saharan cities or states, using the examples of Jenné-jeno, Aksum, and Great Zimbabwe. Finally, I contrast centralized versus household production and propose how to differentiate between these two modes of production in the archaeobotanical record.

Urbanism and Urbanization

The study of what cities are and how and why they formed crosses many disciplinary thresholds, ranging from geography, architecture, sociology, and engineering to anthropology (e.g., Marcus and Sabloff 2005; Ucko et al. 1972). Cities were characterized by a suite of traits identified by Childe (1950:9) including: high population size and density; craft specialization; surplus food production; monumental public architecture; centralized control; several tiers of hierarchy and bureaucracy; and written records. This trait-based approach to urbanism was instrumental to early discussions of the ancient city and its recognition in the archaeological record (Adams 1966; 1981; Childe 1950; Northam 1975; Sjoberg 1960; Trigger 1972; Weber 1962), but it was tailored to the urban and state forms then known from Mesopotamia and Mesoamerica (LaViolette and Fleisher 2005; Smith 2003). The trait-list approach conflates urban phenomena with those of the state, and thus might only be useful in discussions of how cities perform within states. This method of recognizing urban forms was influenced by a cultural evolutionary framework (critiqued in Dunnell 1980).

This list-based characterization has been critiqued by Africanist scholars because it fails to recognize many of Africa's indigenous urban systems (see LaViolette and Fleisher 2005 for a detailed review). Many African cities successfully arose without the hierarchical organization, public monuments, or writing systems traditionally associated with urbanism (Fletcher 1998; McIntosh 2000; McIntosh and McIntosh 1993). Examples of such cities are Jenné-jeno (McIntosh and McIntosh 1980, 1984, 1993), Benin (Connah 1975), and Great Zimbabwe (Garlake 1973; Huffman 1972; Pikirayi 2001).

The momentum in anthropological studies of urban life has shifted towards a more functionalist approach, first recognized by Weber (1962), which defines a city in terms of its "performance of specialized functions in relation to a broader hinterland" (S. McIntosh and R. McIntosh 1993:625, paraphrased from Trigger 1972:577). This shift in perspective enables a more dynamic analysis of cities, and has given rise to new methods of investigating urban systems, such as surveying and excavating hinterland settlements (Schwarz and Falconer 1994); increasing interest in local economies such as craft production (Kramer 1994); and more comprehensive sampling within cities themselves (Kenoyer 2005). More recently, scholarship has focused on viewing cities as they were experienced by <u>all</u> their inhabitants, not merely by those who possessed the power or wealth to influence urban planning and administration (Smith 2003).

Subsistence data are particularly well-suited to address these contemporary concerns in urban archaeology because food remains are common components of household waste deposits, and because specialized food economies were often critical to urban trade economies and site provisioning (Hastorf 1993; Zeder 1991). Moreover, the individuals responsible for food production, namely farmers, agricultural laborers, and food preparers, are typically not considered in urban studies, but their contributions can be made visible through the analysis of plant and animal remains in the archaeological record. Food production systems are powerful manifestations of the urban phenomena that create them, as well as the social relations that maintain them, and as such provide a powerful lens through which to view the multiple layers of urban society (Smith 2003; Zeder 2003). In the next section of this chapter, I discuss theories of how urban food production systems arose and functioned, and how urban food production strategies may be identified through archaeobotany, the study of ancient plant remains.

Urban Food Production in the Archaeological Record

The production of an agricultural surplus for provisioning or trade is often cited as a key feature supporting a city or a state. Food that could be stored and fed to non-food producing elements of society, or traded for necessary resources, must certainly have been important to the success of early cities. Specialized and centrally managed food production systems designed to generate and control food surpluses have been documented for many cities and states around the world (Hastorf 1993; Johnson and Earle 2000; Wittfogel 1957; Zeder 1991). This is particularly clear in Mesopotamia where record keepers documented the grains levied as residential taxes, as well as the distribution of grains and bread as wages, on cuneiform tablets (Damerow 1996).

Zeder's Model of Direct vs. Indirect Access to Animal Resources

In her comprehensive analysis of urban food production Zeder (1991, 2003) described a model of animal use at Tal e Malyan in Iran during the Banesh (3400-2800 B.C.) and Kaftari (2400-1600 B.C.) phases of urban development and florescence. She developed a model of the expected representation and distribution of animal elements at urban sites within states. As seen in Table 2.1, Zeder predicted observable differences to arise when individuals have direct as opposed to indirect access to meat resources.

Recipients of meat provisioning are expected to have access to a limited range of high value meat parts, while animal food producers will be able to access a whole range of animal resources. Inherent in this model is the implication that herders will alter killing patterns to suit the needs of the administration.

Table 2.1. Zeder's model contrasting the herder zooarchaeological profile (direct access to meat) with the urbanite zooarchaeological profile (indirect access to meat) (adapted from 1991: 43, Table 1).

Variables	Direct Access	Indirect Access
Production Goals	Herd maintenance, production for family and exchange	Maximization of meat yield and ease of distribution; standardization
Species diversity	High (including wild)	Low (largely domesticates and wild prestige resources)
Selection for	Higher reproductive rates (ie, goats and sheep)	Higher meat yield (ie, cattle and sheep)
Age/Sex diversity	High (ie, males between .5-2 yrs)	Low (ie, males between 2-3 yrs)
Elemental representation	Even, wide distribution	Restricted to select parts, limited distribution

Zeder (1991) found interesting patterns, some of which contradict the expectation that urban administration exerted control over herding practices. Species distribution patterns were generally more even across space and time than predicted, possibly due to the narrow range of animals (cows, sheep, goats, donkeys, and infrequently pigs) available in Bronze Age Mesopotamia (Zeder 1991: 246). Narrow age distributions were particularly notable in administrative or craft production centers (Zeder 1991: 247), but did not always correspond with meat maximization. Indirect provisioning was most evident through analysis of the parts represented: medium sized animal (sheep and goat) limb bones proved the most common elements in specialized urban contexts (Zeder 1991: 247). Moreover, ovi-caprid head elements were found in abundance in one context, suggesting its function as a processing center for incoming animals. However, there was some evidence for whole animals being available to households in administrative quarters, suggesting that animals were occasionally distributed in their entirety (Zeder 1991: 248). Finally, herd management strategies were largely obscured in the Malyan zooarchaeological assemblage, which was predominantly urban and thus did not represent village contexts most likely to contain patterns of direct decision making (Zeder 1991: 249). Zeder's model highlights limitations of idealized models and shows that urban food production may incorporate elements of both direct and indirect access.

Producer/Consumer Models of Agricultural Production

Plant food production modeling follows several of the human behavioral expectations laid out in Zeder's model, but necessarily take into account the natural properties of plants, particularly domesticated grain taxa. Like zooarchaeologists, archaeobotanists have derived models of food producers (direct access) versus food consumers (indirect access) based on the relative proportion of food products to production waste as observed in ethnographic examples (Dennell 1974, 1976; Hillman 1973, 1981, 1984; G.Jones 1984; M.Jones 1985). Ethnographic observation of wheat and barley processing practices revealed a series of stages common to most production systems (Hillman 1973, 1981; G. Jones 1984). In general, it was observed that the early stages of crop processing, including threshing, first winnowing, and coarse sieving, resulted in the separation and discard of small weeds, straw (culm) and the rachises of free-threshing wheat. The later stages of crop processing largely serve to remove the glume portions from the grains through pounding, fine sieving and further winnowing, and result in the discard of fine chaff fragments and larger weed seeds.

The use and utility of producer-consumer models in archaeobotany has come under scrutiny. Despite their dependence upon a similar processing sequence, M. Jones (1985) and Hillman (1981, 1984) identify producer and consumer archaeological assemblages differently based upon the behaviors surrounding discard patterns (Stevens 2003). Using an economic model of relative abundance and scarcity, Jones assumes that producer sites will contain a higher proportion of grain to chaff because consumer sites will inordinately value grain and thus ensure that it doesn't become discarded. Hillman's model, on the other hand, makes the assumption that grain will be transported in the form in which it is stored, and that consumer sites will show evidence of post-storage processing (light chaff elements, big weeds) while producer sites will be characterized by the predominance of pre-storage processing residue (straw, free-threshing rachises, small weeds). As noted by Stevens (2003), M. Jones and Hillman interpret the same data in opposite ways, which limits the interpretive value of the producer/consumer archaeobotanical signatures.

Van der Veen (1991) reported difficulty applying either Hillman's or M. Jones' models to a Roman archaeobotanical assemblage in Britain, finding that both an administrative center and a farming complex demonstrated moderate producer signatures archaeobotanically. Bases on this example, Van der Veen suggests that the producer/consumer dichotomy be expanded to four points along a continuum of production and consumption, including subsistence production, surplus production, small consumer sites, and large urban complexes.

Recent critiques have indicated that producer-consumer models often have little application outside of their original study areas, due to the effects of cultigen morphology, rainfall regime, and social organization on crop processing and storage practices around the world (Smith 2001; Stevens 2003). Reddy (1997) examined the stages of non-mechanized sorghum processing in India to model agricultural production and trade in Harappan communities in Gujarat, India. Stevens (2003) suggests that the household/communal processing model is a more accurate interpretation of the archaeobotanical signatures outlined in the production/consumption models of Hillman, Jones, and van der Veen discussed above. Household-based crop production in Stevens' model is characterized by a low proportion of pre-storage processing compared to the daily processing required to clean grains for food preparation.

Another line of critique challenges how closely archaeological samples represent the activities that created them (Fuller et al. 2005; Miksicek 1987). Most archaeobotanists recover charred plant remains (Fuller et al. 2005; Reddy 1997). Yet not all food products fare equally well when placed in contact with fire, and the lighter chaff elements are more fragile and less likely to remain identifiable in the archaeological record. Ironically, production signatures rely most heavily on the botanical components least likely to become carbonized: the culm fragments and rachises. Ultimately, production/consumption models are most informative when combined with other archaeological or environmental data to characterize the food production strategies in existence at a given archaeological site (van der Veen, 1991).

Recognizing Agricultural Centralization

Archaeological signatures of food production could be obtained by examining structures, cultural materials, and plant remains in both domestic and communal archaeological contexts (see Table 2.2). The centralized food production systems characteristic of hierarchical administrations are best recognized by the presence of centralized food production facilities for communal labor. Alternatively, in a household mode of production, food production activities take place predominantly in or around the house, with household members or relatives serving as the primary labor source. Similarly, in a centralized system, storage facilities would have been large and under the control of the administration, whereas storage facilities in household-based production systems would be located in or adjacent to individual households.

Plant-waste products generated by processing before storage would most likely be concentrated in specialized production centers in the case of centralized production, but, in the case of household-based production, processing waste would be dispersed among households and middens throughout the community.

Table 2.2. Characteristics of centralized food production system versus household based food production.

Variable	Centralized Food Production	Household Food Production
Grain storage	A few large protected stores located in urban contexts	Many small dispersed stores located in domestic contexts
Pre-storage processing	Large-scale, centralized	Small-scale, widely dispersed
Control of stores	Administration	Household head or members
Location of trade goods	Concentrated among the administration and elite	Dispersed among domestic contexts

Moreover, the communal labor involved in centralized production would likely result in cleaner assemblages of stored grain, while household stores could afford to contain more chaff because household labor would be available for daily processing prior to meal preparation (Stevens 2003). Finally, goods associated with the exchange of food products would also have different patterns of distribution depending upon the level of centralized control. Within a centralized food production system, only the administrative element of society could exchange food products for foreign goods. Conversely, according to the household mode of production, each household that could mobilize surplus food products for market exchange could possess foreign trade items.

Differentiating between centralized versus household production systems is contingent upon detailed knowledge of the crops in question and a thorough archaeobotanical analysis of the archaeological contexts. The production of "cash" crops in addition to subsistence crops might occur in communities specializing in long-distance trade, and would be signaled archaeobotanically by the restriction of certain plant taxa within administrative storage or processing contexts. Among the different grain taxa, morphological differences may mean different crop processing stages. For example, the large condensed spike of pearl millet can be readily collected by hand to avoid including weed species while harvesting (Acland 1971:28), thus eliminating the need to remove weed seeds by winnowing later in the process. The harvesting of wheat and barley, however, is most efficiently done by sickle or plough, methods which include weeds in with the grain plants and thus necessitate removal of weed seeds via several phases of winnowing (D'Andrea and Mitiku Haile 2002; Hillman 1973, 1984; Jones 1984). Therefore, it is instructive to consider African models of agricultural and urbanization.

Urban Food Production in Africa

Pathways to both food production and urbanism appear to follow different trajectories in Africa than among original models derived from Near Eastern examples. It is therefore reasonable to predict that that urban food production patterns in Africa may also differ from the classic Near Eastern model, due to differences in the plants and animals that form the subsistence base as well as in the urban forms that supported, and were supported by, food production systems. In this section I present several case studies of urban food production in Sub-Saharan Africa, focusing on early cities from which archaeobotanical data have been collected. Readers should note that this sampling is limited by the availability of urban food production data in the Africanist literature, and that the examples provided show marked heterogeneity in the methods and scale of analysis employed. Yet, excellent work has begun in this area, as will be seen in discussions below of urban food production at Jenné-Jenno, Aksum, and Great Zimbabwe.

*Aksum (Highland Ethiopia; Sixth and Seventh centuries A.D.)*This capital of the Aksumite Empire in Ethiopia (see Figure 2.2 for map of African sites mentioned here) has yielded important information about early urban plant food production in Africa (Boardman 1999, 2000). Data regarding relative abundance of plant taxa before and during the Axumite empire are summarized in Table 2.3 below. Interesting trends can be noted, including a large net increase in the variety of taxa used, and the fact that no cultivars present during Pre-Aksumite times disappear during the Aksumite empire. This appears to contradict expectations that urban food production will lead to taxonomic specialization. Moreover, Near Eastern crop species are favored during the pre-Aksumite period, while all African grains except the indigenous tef are found only later, during the Aksumite period.

Among the cereals, grains were found to far outnumber chaff elements, leading Boardman to suggest that grains were processed away from the excavated structures and middens during Axumite times, in contexts where they were unlikely to become charred or otherwise deposited archaeologically (Boardman 1999). This could include areas of the city yet to be excavated, or contexts outside of the urban milieu. The few contexts that do show evidence for crop processing are likely the result of Aksumite residents producing their own food, while the dominance of grains at most deposits likely signals indirect access to grains through provisioning or market trade. Boardman suggests that plants in the city and empire of Aksum played significant roles in both domestic and market economies:

Figure 2.2. Map showing African archaeological sites and regions mentioned in the chapter: (1) Jenné-jeno; (2) Benin City; (3) Aksum; (4) Great Zimbabwe; (5) Pemba Island.

> [T]here may have been a tier of agricultural production units above the family farm, which may have been geared towards meeting local urban needs as well as tribute payments, and possible even towards markets further afield. It is likely that farmers within the immediate hinterland of Aksum had to supply tribute payments, although the relative roles of trade and tribute in the expanding agricultural system are not yet clearly understood....[G]rain has traditionally been a high bulk, low value product, most likely to have been exchanged over short distances, usually for other agricultural products. The apparent introduction during the Aksumite period of new cultigens, such as oil and fiber crops, fruits and vegetables, is noteworthy in this respect, as these are often grown primarily for market. (Boardman 1999:45)

In addition to the archaeobotanical data, some rooms bear evidence of specialization and the provisioning of certain members of the society at Aksum. Some structures appear not to have had domestic functions, and one such context produced an abundance of grinding stones and plants that produce oil, noog and cotton (Phillipson et al. 1996).

Cain's (2000) study of the faunal remains at Aksum reports data that are directly applicable to the urban food production question. Significantly, domesticated bovids, particularly cattle, were the dominant taxon (cattle represented 85% of identifiable elements). Cattle were likely valued for their secondary products and for their labor, such as in the *maresha* plow system, in addition to their meat value. During Aksumite times, two different areas of urban occupation can be discerned at Aksum, including the peripheral D-site, characterized by evidence of agricultural production, and the central K-site, an apparent locus of craft industry and a net consumer of both cleaned grains and butchered meat (Boardman 1999, Cain 2000:154). At D-site, a broader range of animal parts is observed, as are older animals, which suggests a domestic production pattern in which animals are retained past their prime meat-yielding stage in order to be used for work, breeding, or milk production (Cain 2000:152-153). This contrasts with K-site, where the animal parts tend to be restricted, and the dominance of young cattle is consistent with the practice of butchering for meat (Cain 2000:154).

Evidence from animal bones at Aksum suggests that while animal production itself did not undergo specialization, it is possible that specialization occurred, evidenced by the butchery and redistribution of animal parts among the population (Cain 2000: 148). Similar to the patterns noted by Zeder (1991, 1999), urbanization and state formation at Aksum did not lead to radical alterations of herding strategy, as indicated by the presence of older animal bones as well as the young adult males predicted for meat-maximizing pastoral production. Moreover, the zooarchaeological data indicate that redistribution could have been a market event as opposed to an administrative function; Aksum minted its own coinage, and conceivably Aksumites could have purchased their meat with money or trade in kind (Cain 2000:154-155). Thus, choice of meat species, cut, and quantity at Aksum, and particularly K-site, could have been more a function of market selection, income level, and/or culinary preference, than a dictum of administrative redistribution. This finding accords well with Boardman's (1999, 2000) reports of plant species

Table 2.3. Relative abundance of cultivated plants recovered from pre-Aksumite (5th century B.C.) and Aksumite (6th -7th centuries A.D.) deposits. From Boardman 1999:142. Scale: (-) absent, (?) questionable, (+) rare , (++) common, and (+++) abundant.

Cultivars		Pre-Aksumite	Aksumite
Cereals	2-row barley (*Hordeum distichon*)	+++	+++
	tef (cultivated and wild) (*Eragrostis* spp.)	++	+++
	emmer wheat (*Triticum dicoccum*)	+	++
	bread wheat (*Triticum aestivum*)	-	+
	UNID wheat (*Triticum* sp.)	+	++
	finger millet (*Eleusine coracana*)	-	+
	sorghum (*Sorghum bicolor*)	-	+
	oat (*Avena sativa*)	+	++
Pulses	lentil (*Lens culinaris*)	?	++
	pea (*Pisum* sp.)	-	+
	grass pea (*Lathyrus sativum*)	-	+
	chick pea (*Cicer arietinum*)	-	+
Oil/fibre	flax (*Linum usitatissimum*)	+++	+++
	noog (*Guizotia abyssinica*)	?	++
	cotton (*Gossypium* sp.)	?	+
Fruit/ Vegetable	mustard family (*Brassica* sp.)	+	++
	grape (*Vitis vinifera*)	-	+
	bottle gourd (*Lagenaria siceraria*)	-	+
	east African cordia (*Cordia africana*)	-	+
	UNID nutshell	+	+

expansion, particularly items such as fruits and vegetables that are easily brought to and exchanged at market.

Jenné-Jeno (Inland Niger Delta, Mali; 250 B.C. - A.D. 1400)
The fertile environment of the Inland Niger Delta (IND) bears streams that facilitate transportation, fishing, and African rice cultivation, the latter through yearly cycles of inundation. This powerful combination of mobility and high subsistence yields is credited with supporting the rise of Jenné-jeno and other "urban clusters" in this region by A.D. 400 (McIntosh and McIntosh 2003). Intensive archaeological work has proved that these cities were not the result of later North African trade economies as previously thought, but rather that African settlements and societies preceded, but later permitted participation in, trans-Saharan trade (Connah 2001; LaViolette and Fleisher 2005). Jenné-Jeno itself has been the subject of intense excavation and multi-disciplinary investigation, and has yielded some of the only archaeobotanical remains for this region (McIntosh 1995a; McIntosh and McIntosh 1980).

S. McIntosh (1995:348-353) reports the results of 126 ten-liter flotation samples analyzed by Jack Harlan, John Sscheuring, Youssouf Boré, and Abdoulaye Sow. From the earliest occupation at Jenné-jeno, the inhabitants used domestic African rice (*Oryza glaberrima*), pearl millet (*Pennisetum glaucum*), and sorghum (*Sorghum bicolor* ssp. *bicolor*), taking advantage of both the floodplain basin (rice) and the drier floodplain levees (millets). Thus, the colonizers of the site were likely already familiar with the local agricultural conditions and arrived with a suitable suite of crops in hand. Among domesticates, African rice was the most common by far, along with sedge nutlets and browntop millet grains, discussed below.

Perhaps the most surprising element of the archaeobotanical record at Jenné-Jeno is the presence of the wild, possibly cultivated, grain browntop millet (*Brachiaria ramosa*), which is found in virtually every sample and constitutes the vast majority of plant remains in some samples (analyses were presented as presence/absence only, but occasionally relative abundance was noted). Browntop millet, along other wild grasses found at Jenné-jeno such as with barnyard or Japanese millet (*Echinocloa colona*), wild African rice (*Oryza barthii*), and a type of panic grass (*Panicum laetum*), are all collected, stored, and brought to market as valued goods in this region today (Boré 1983 in S. McIntosh 1995:351). The Jenné-jeno samples also contained many different wild grass grains at low abundance.
Interestingly, the botanical assemblage remains remarkably stable over time, suggesting that the site was founded on a particularly well-adapted subsistence economy that was able to feed an increasing population through extensification. This parallels the relative taxonomic stability observed in the zooarchaeological (MacDonald 1995) and ichthyofaunal (van Neer 1995) records from Jenné-jeno, which demonstrate a reliance on domesticated bovids and wild game as well as a wide variety of fish caught from rivers, seasonal pools, and swamps. Thus, the rise of urbanism in and around Jenné-jeno appears to have been based on an already productive and wide subsistence base, and did not result in a specialization of the economy through elimination of particular taxa, or intensification of agricultural production. Connah (1985) argues that use of recessional cultivation along the floodplains constitutes a form of intensification in its own right, and that "the urban growth in the Inland Niger Delta was supported by one of the relatively rare examples of African intensive agriculture" (Connah 2001:131). Considering that floodplain cultivars such as rice and browntop millet were economic staples even before urban expansion, however, it is difficult to determine the relative contributions of extensification and intensification to supplying the urban food base. Detailed analysis of the redistribution of food products awaits quantification, particularly of the botanical remains, however McIntosh hypothesizes that:

> because transport costs are low on the floodplain and the social benefits of 'going to market' high for rural dwellers, it is more likely that the produce of distant farms was regularly transported by pirogue [dugout canoe] into Jenné-jeno, which would have been an important market center for staple commodities. A growing importance of local trade in consumables probably accounts for the lack of evidence for intensification of subsistence production at Jenné-jeno. (McIntosh (1995:378)

Great Zimbabwe and the Zimbabwe Tradition (Zimbabwe; A.D. 1000-1500)
The Zimbabwe plateau was home to a state built on mixed farming and mining practices, and characterized by zimbabwe sites bearing granite enclosures built using drystone technology (Pikirayi 2001). These enclosures served as shelters for other structures or features, including granaries, groups of earth and thatch huts, or ritual items (Connah 2001:231-232). The largest and most impressive of these sites is Great Zimbabwe, whose Great Enclosure is an impressive 84 m in diameter, five meters thick and 10 meters high. At the height of its power (A.D.1250-1450) Great Zimbabwe may have housed upwards of 10,000 residents (Sinclair 1987:114). While it is debated whether Great Zimbabwe and the other postulated state capital (Mapungubwe) are cities in their own right, their "scale, monumental public architecture, long distance trade relations with the Swahili coast, elite residences [and] sculptural traditions related to spiritual power" attest to the importance and power of these centers at the floruit of their existence (LaViolette and Fleisher 2005:338).

Both faunal and floral resources have been reported for Great Zimbabwe, as well as some other sites within the region. Remains of the animal economy at Great Zimbabwe, and among other related sites, indicate a heavy reliance upon domesticated bovids, and cattle in particular. For example, among the fauna recovered from a midden down slope from the Hill Ruin at Great Zimbabwe, Brain (1974) found that 98.6% of identifiable materials were attributable to cattle, but that sheep/goat, dog, and some wild taxa were also represented. Garlake (1978) has suggested that a specialist transhumant cattle economy evolved which allowed movement between the lowlands and the plateau to take advantage of seasonal tsetse-free pastures. Even without such specialization, however, cattle would have been remarkable sources of meat and wealth with which to overcome seasonal shortages.

Botanical research has been well integrated into studies of the Zimbabwe past, and recently synthesized by a major contributor, Jimmy Jonsson (1998). Early hunter-gatherer and transitional farming sites (7000-500 B.P.) on the Zimbabwe plateau demonstrate reliance upon wild fruits and grasses from a variety of habitats, and later the domesticated groundnut. Early and late farming communities during the present era continued to rely on wild cultivated grasses, however domesticates were introduced through time. For example, at Mwenezi Farm between A.D. 700 and 900, the botanical record was dominated by cultivated wild grasses, including: guinea grass (*Panicum maximum*), signalgrass (*Brachiaria deflexa*), sweet signalgrass (*Brachiaria eruciformis*), crowfoot grass (*Dactylotenium aegyptium*), goosegrass (*Eleusine indica*), Lehmann's lovegrass (*Eragrostis lehmannia* var. *lehmannia*), and whorled dropseed (*Sporobolus pyramidalis*). By A.D. 500, domestic finger millet makes its first appearance at the site of Kadzi (Jonsson 1998:108), and domestic sorghum is found by the seventh century at Leopard's Kopje main krall (Huffman 1974a: Table 15, reported in Jonsson 1998:63). Domestic pearl millet has a later distribution among Zimbabwe tradition sites (after the fourteenth century), but is present at sites in southern Africa from the third to fourth centuries A.D. (Silver Leaves site, South Africa; Klapwijk 1974).

Jonsson's (1998) treatment of plants in the Zimbabwe past is tuned towards understanding how early hunter-gatherers, and, later, agro-pastoralists weathered Holocene climatic variability through shifting resource utilization. Among the different wild and cultivated grasses there are different habitat preferences, leading Jonsson to track changes in the local environment through time. While this is a fascinating and valuable contribution in its own right, there is a lack of high-resolution botanical data with which to construct the role of plant foods in the rise of state and city in Zimbabwe. This is not Jonsson's problem alone, but rather one that stems from some of the biases in archaeological sampling in this region (Connah 2001:248). One major deficit lies in the focus on stone structures and the activities within them, which are likely to pertain to elite lifeways. Yet we know little about foods consumed in daga (earthen) structures outside of the stone settlement areas at Great Zimbabwe, or at smaller sites. Even at the stone sites with the best botanical representation, wild grasses are still present, and in some cases, dominant, thus the persistence of a wide subsistence base must have been necessary to support the local and trade economies during the development of the state. This broad subsistence base, in addition to keeping cattle for food and wealth, likely enabled the Zimbabwe state to bridge the transition from a more humid to a drier climate at around A.D. 1300 (Jonsson 1998:114).

Summary

While the three sites discussed above differ in many ways, including time period, scale, and environment, several important trends emerge from a consideration of their food production strategies. Most striking is the expansion or maintenance of a wide subsistence base, including wild plants, domesticated plants and bovids, and, at Jenné-Jeno in particular, wild animals and a variety of fish species. This contradicts the specialization of food economy predicted by classic models of urbanization, where standardization is seen as a key element within the food distribution system. A wide subsistence base likely served to guard against unpredictable resource stresses from droughts, pestilence, and other food crises. A second commonality among the African examples is the importance of the market in food redistribution, which seemed particularly central to the food production systems at Aksum and Jenné-jeno. There, both cities are envisioned as important market centers to which farmers and herders, and presumably craft specialists, brought goods for sale. This system likely served both the producer and the consumer well: market produce gave an income or trade opportunity to rural dwellers, and it provided city folk with easy access to necessary goods and services, like grain, vegetables, meat, cloth, pottery, iron tools, and the communication network such a forum provides. The presence of such a market system reduces the likelihood that urban or state administration was involved in much food redistribution activity. Thus, the centralization of food production expected under the classic urban model does not appear to have been necessary to provide the urban core with adequate food supplies in the African cities described above. It will be interesting to see if future work bears out similar patterning, including Louis Champion's archaeobotanical work in Benin (Champion and Fuller 2015).

Conclusions

While classic urban plant food production models derived from "primary" cities of the Near East and Mesoamerica will have to be altered to suit the plant crops under consideration and the likely goals of surplus food production (ie, provisioning versus trade), some general trends pertain. These include specialization of plant taxa, centralization of storage and distribution, and organization of communal labor for pre-storage processing. While urban food production systems in the Near East, such as Tal e Malyan, appear to have become standardized to administrative requirements in terms of the distribution of goods, herding practices appeared largely unchanged and outside of administrative control during urbanization (Zeder 1991). Plant food production systems have been largely conceptualized under the consumer-producer dichotomy (Hillman 1984; Jones 1985; van der Veen 1991), but newer approaches have emphasized the role of labor and social organization in determining plant processing strategies and the distribution of plant remains archaeologically (Stevens 2003). Urban systems are expected to exert control over the plant food resources necessary for provisioning and trade, resulting in the specialization of food or cash crops accordingly. Administrative control is also expected to lead to the organization of communal labor for centralized processing and distribution and the standardization of plant food products by type as well as amount.

These expectations of urban food production require testing against the archaeological and historic records to refine them and to assess their applicability across diverse urban systems. Clearly, the African examples of Jenné-jeno, Aksum, and Great Zimbabwe suggest alternative modes of urban food production. How did Swahili food production change with the rise of stonetowns? Swahili communities on the Eastern coast of Africa participated in extensive trade network that connected Africa's eastern interior with Arabia, South Asia, and China. This study sampled four archaeological sites on Pemba Island, Tanzania, to document plant use and production in Swahili towns and villages dating between A.D. 600 and 1500. In the next chapter, I place the ancient Swahili in geographic, economic, and socio-political context through a survey of the archaeological, ethnohistoric, and linguistic evidence.

CHAPTER 3: PEMBA IN THE SWAHILI WORLD: NATURAL AND CULTURAL CONTEXTS OF SWAHILI LIFE

In this chapter I provide an introduction to the Swahili and their environs. I summarize the climate patterns, sediments, and vegetation zones in eastern Africa to construct a view of the natural environment encountered by the Swahili and the inland communities with whom they interacted. I also situate Pemba within the Swahili cultural framework as is known from past and ongoing excavations as well as ethnographic and historic accounts. I summarize what is currently known about the four settlements archaeobotanically surveyed for this work: Tumbe, Kimimba, Chwaka, and Kaliwa. Finally, I propose a model of how Swahili urban food production may have operated in the past using archaeological, ethnohistoric, and linguistic data. In Chapter 7, this urban food production model will be used to test the archaeobotanical data presented in Chapters 5 and 6.

Eastern African Coast in Geographic Perspective

Holocene Climate of Eastern Africa

Understanding the climatic conditions that prevailed during the Holocene in eastern Africa is vital to interpreting the events witnessed in the archaeological record, particularly among settled farming communities such as the Swahili, which likely relied on rain-fed agriculture before wet rice cultivation techniques were introduced. While Holocene climatic patterns are often generalized in terms of a gradual shift from conditions prevailing during the last glacial maximum (18 ka B.P.), to a global thermal maximum at 6k B.P., to the prevailing climate of today, regional data show that several abrupt events occurred during the Holocene (2005; Hassan 1997; Russell and Johnson 2005; Street-Perrott and Perrott 1990).

Holocene climate records in eastern Africa derive largely from the analysis of sediment cores from the area's highland lakes. Closed lake systems show exaggerated responses to environmental conditions and thus serve as sensitive proxy indicators of climate change (Street-Perrot and Perrott 1990). By the beginning of the Holocene (12-10 ka B.P.) the conditions that prevailed in tropical Africa were wetter than at any other time since (Hamilton 1982:245-253). Severe drought conditions appear to have hit eastern Africa at 8700-8500 cal B.P. and periodically during the latter third of the Holocene. Both Mazuka et al. (2004) and Grab et al. (2005) present data that show an abrupt change to more arid prevailing conditions in eastern Africa at approximately 8600 cal B.P.. Using geochemical analyses of biological silica and magnesium levels at Lake Edward, Russell and Johnson (2005) detected drought signals at: 4100 cal B.P., 2000 cal B.P., 1500 cal B.P., and a widespread drought occurring between 750 and 950 cal B.P. that would have impacted all of eastern Africa. This latter signal is found in several studies throughout the region. Diatom records from Lake Victoria (Stager et al. 2005) show low lake levels between 820 and 620 cal B.P.. Filippi and Talbot (2005) infer low lake levels at 700 cal B.P. in northern Lake Malawi based on organic content and isotopes from lake core sediments.

Meteorology

Much of the modern annual weather cycle along the eastern coast results from two monsoon cycles (see Figure 3.1). The Northeastern monsoon brings hot, moist air from the Arabian Peninsula and directs the Indian Ocean currents towards Africa from November to March, bringing with it sailing vessels from South and Southwest Asia. The southern monsoon brings cool, moist air from the Indian ocean to the coast and directs wind north up the coast, first at an easterly angle (Southeast monsoons, May-June) and then to the northwest towards the Arabian peninsula (Southwest monsoons, July), This bidirectional wind pattern over the course of the year has an enormous impact on the natural and social calendar of the Swahili, bringing both rain and sailing opportunities for long-distance trade vessels (Horton and Middleton 2000; Pritchard 1975).

The timing, duration, and strength of the two annual rain seasons are critical for the success of cultivation, particularly the rain-fed fields and gardens likely used by the Swahili in the past. Recent investigations into the prediction of the rainfall in eastern Africa has shown that sea-surface temperature is highly correlated with the strength of rains, particularly the November-December rains brought by the North-Eastern monsoons (Mwale and Gar 2005). Low end-of-year rainfall is associated with a cool sea-surface in the Brazilian Current and the southern Indian Ocean, but is associated with warm temperatures in the northern Indian Ocean in the preceding monsoon cycle (Mwale and Gar 2005).

Landforms and Sediments

There is a general gradation of altitude and rainfall in eastern African from east to west (see Figures 3.2-3.4). Much of the eastern African highlands are covered by volcanic beds that support intensive cultivation due to the rich, well-drained volcanic soils and high (1000 mm) annual rainfall (Pritchard 1975:11, 56). The drier plateau behind the coast, however, is formed from sedimentary deposits and bears impoverished, poorly drained soil types, such as Black Cotton, that are difficult to cultivate and do not support permanent structures (Pritchard 1975:34).

Much of the coast lies on marine deposits from a major period of Miocene transgression, including various limestone formations and beds of sands, silts, and clays (Schlüter 1997:243-245). At the coastal edge, and on Zanzibar Island[1], Quaternary and Holocene reef deposits dominate and, when outcropped, render the soils unsuitable for intensive agriculture, such as on much of Zanzibar's eastern half. On Pemba and Mafia Islands, older marine and fluviatile deposits from the Miocene dominate, and "broad swampy aggraded valleys are indicative of late subsidence" (see Figure 3.5; Schlüter 1997:251). The coastal sediments are typically characterized as sandy loams (Kusimba 1999:73). Soil content and quality varies along the coast, resulting in an uneven and patchy distribution of lands suitable for cultivation (Fleisher 2003:13-14).

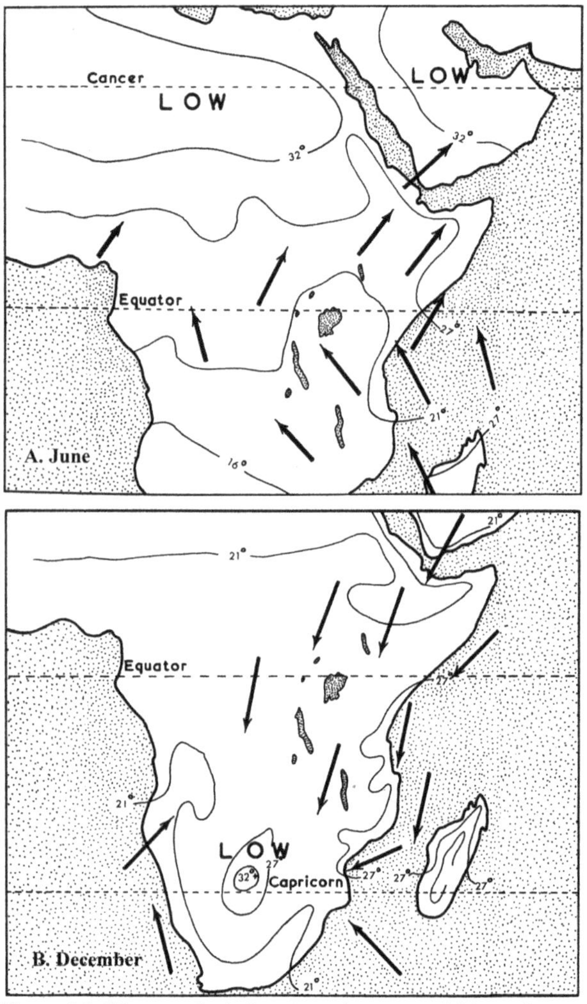

Figure 3.1. Monsoonal wind patterns in eastern Africa today: (a) Southern winds predominate in June; (b) Northeast winds predominate in December. Adapted from Pritchard (1975:27-28, Figures 12-13).

Landscapes

Swahili communities occupied a long but narrow band of the African coastline, extending 2500 km from Mogadishu in present-day southern Somalia (3° N) to Sofala in Mozambique (20° S), and including the offshore islands of Pemba, Zanzibar, Kilwa, Comoros Islands, and the northwestern tip of Madagascar. Depending as they did upon the Indian Ocean for trade, communication, and fishing, the Swahili rarely settled beyond 20 km inland from the coast, possibly because the narrow rivers did not allow boat passage inland. While the limitation of the Swahili to the coast is often overstated and

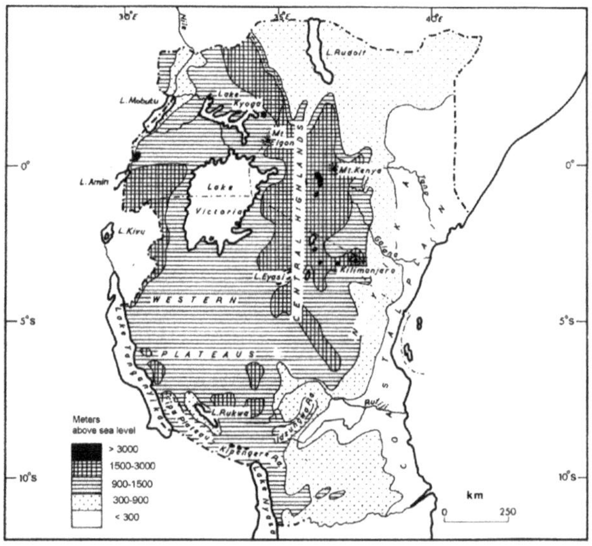

Figure 3.2. Landforms and elevations in eastern Africa. Adapted from Pritchard 1975:14, Figure 6).

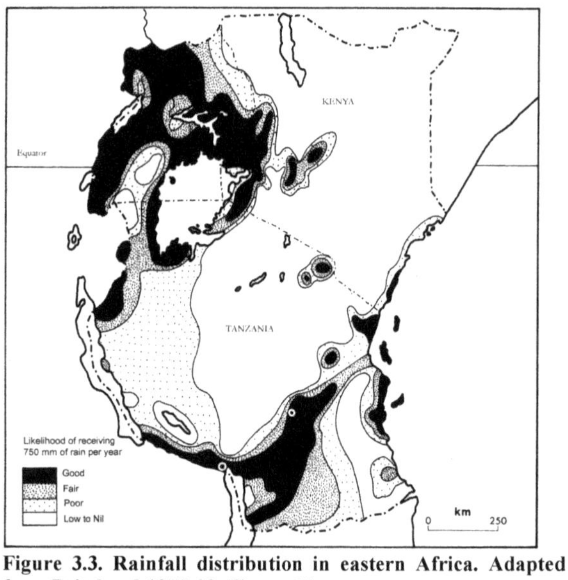

Figure 3.3. Rainfall distribution in eastern Africa. Adapted from Pritchard 1975:29, Figure 15.

[1] Zanzibar is the current political entity encompassing the Tanzanian islands of Pemba and Zanzibar, the latter of which is known locally as Unguja. While the politics of the past 200 years has paired Pemba with Zanzibar, in the past Pemba was more closely associated with Mombasa and the surrounding coast than it was Zanzibar Island.

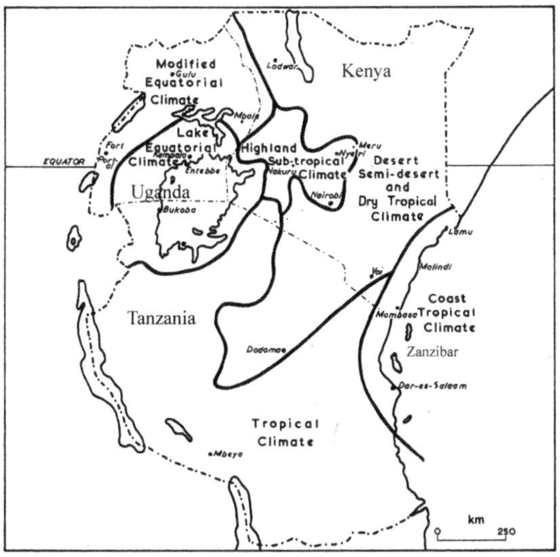

Figure 3.4. Climate regimes of eastern Africa. Adapted from Pritchard 1975:30.

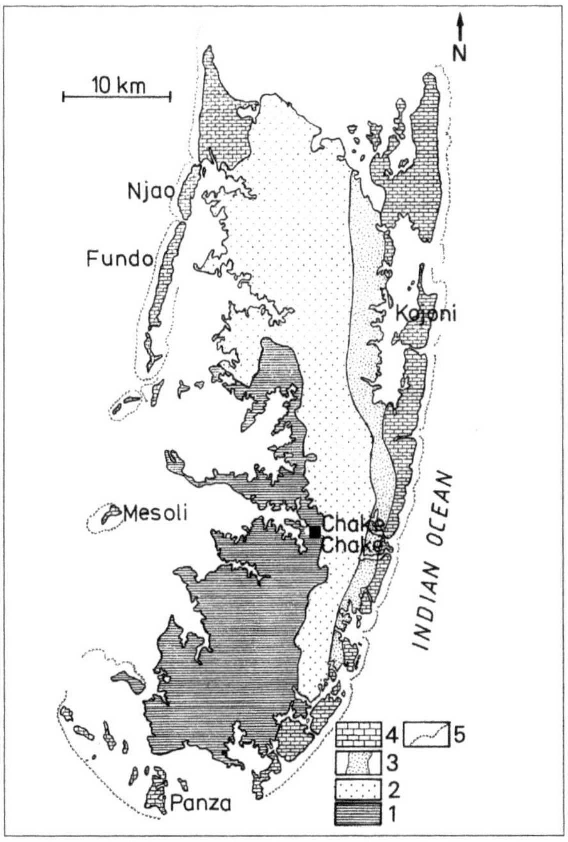

Figure 3.5. Geological map of Pemba Island. Adapted from Schlüter 1997:249, Figure 102: (1) Lower Miocene Chake Chake beds (marine clays, sands, fossiliferous); (2) Miocene/Pliocene Weti beds (fluviatile/estuarine); (3) Pliocene Vitongoge beds; (4) Pleistocene Azanian limestone (raised reef); (5) Holocene fringing reef.

under-theorized (Fleisher 2003:15), recent research has brought Swahili coastal-hinterland interactions into focus, as will be discussed below (e.g., Abungu 1989; Chami 1994).

Despite the common low altitude maritime location and bidirectional monsoon winds, the local environments supporting Swahili communities differ along the coast from north to south due to the effects of latitude, geology, and the movement of the Inter-Tropical Convergence Zone [2]. This subtle heterogeneity results in a differentiation of resources along the coast. Six general areas of coastline can be identified, from north to south (see Figure 3.6): Benadir, Visiwani, Nyali, Mrima, Mgao, and Kerimba (Fleisher 2003: 11-13; Horton and Middleton 2000:6-10). The Benadir coast, located north from the Somali-Kenya border, is markedly arid, particularly north of Mogadishu and inland from the coast. The moist Visiwani coast of northern Kenya supports the coral reefs and mangrove swamps associated with the Swahili from the earliest texts, and receives increasing amounts of rainfall to the south. Offshore coral reefs and islands (the Lamu archipelago) provide natural breaks and good harbors conducive to boat travel. Semi-arid woodlands to the interior of the Visiwani coast are more suited to pastoralism than agricultural pursuits. South of the delta of the Tana river to the border with Tanzania, the Nyali coast is somewhat more fertile than the Visiwani owing to higher rainfall, including the coastal hinterland. The Mrima coast, extending from northern Tanzania to the Rufiji delta, boasts the highest rainfall and supports the most intensive agriculture in this region. The three offshore islands of Pemba, Zanzibar and Mafia offer excellent harbors, as do the numerous creeks and beaches. Mangrove forests abound, and the offshore coral reef soils in the north, while plains support coastal forests in the south by the Rufiji river. Inland, the Usambara and Pare mountains bear fertile soils in the north, while plains support coastal forests in the south by the Rufiji river. The Mgao coast of southern Tanzania to the Mozambican border has few safe harbors excepting the famous Kilwa Kisiwani. The Mgao supports the most southerly extent of the coral reef system, and bears much less fertile land than the Mrima and Nyali coasts. Finally, the Kerimba coast represents the southernmost Swahili presence; however, with no coral reef and few mangrove swamps, this area bears little resemblance to its counterparts north of the border of Mozambique.

Thus, areas suited to watercraft, fishing, farming, and the harvesting of mangroves for poles and charcoal are distributed unevenly along the coast. Note that Pemba lies in the Mrima - this portion of the coast is best suited for agriculture, but also supports mangrove swamps and quiet harbors. The uneven distribution of resources along the littoral could have been ameliorated by contact and sharing among communities through water transportation (Fleisher 2003:14). This may also explain in part

[2] The Intertropical Convergence Zone is a band of low pressure belted around the globe where the northeast and southeast trade winds meet. This zone may fluctuate north or south of the equator depending upon latitudinal shifts in the trade wind patterns, bringing moisture with it as it moves.

how and why cultural patterns came to be shared and maintained along this extensive shoreline.

Vegetation

The cultural distribution of Swahili settlements along the coast broadly corresponds to the geographic extent of the Zanzibar coastal Inhambane vegetation zone (White 1983), although the homogeneity of this region has been overstated (Dharani 2002). Patches of mangrove forests along the coastline would have supported saline-tolerant trees and shrubs in protected bays or riverine estuaries (see Table 3.1 for a list of taxa common in mangrove forests).

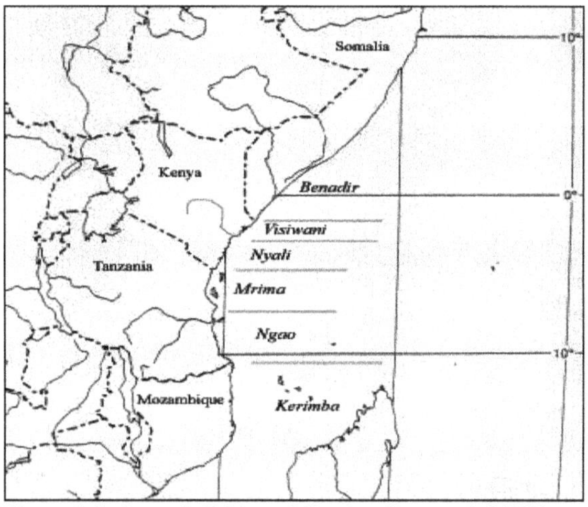

Figure 3.6. Coastal divisions following Horton & Middleton (2000:6-7, Map 1.1) and Fleisher (2003).

Tropical coastal forest, where it existed, extended only a short distance inland, before drier lands reduced vegetation to miombo-type deciduous woodlands (dominated by Caesalpinoid trees, such as *Brachystegia*), shrublands, and grasslands (Dharani 2002:5). Today, much of the land is intensively cultivated where permitted, and cash crop plantations (coffee, sisal, pineapple, tea, cotton) have replaced subsistence farmsteads in many regions, particularly on the drier portions of the mainland plateau where sisal (introduced from the Yucatan in 1893) and pineapple are among the few cultivable crops (Fitzgerald 1945:228).

The extant portions of forest in Pemba (Ngezi forest) and Zanzibar (Jozani forest) are protected. These areas exhibit high taxonomic and plant form diversity, characterized by tall trees [3], cycads, pteridophytes, palms, orchids, lianas and climbers[4], mangrove taxa (see table 3.1 above), shrubs, herbs, heath (*Philippia*), marsh and freshwater taxa (*Nymphoides*, *Nymphaea*) (Beentje 1990; Koenders 1992). Ngezi forest on Pemba Island contains wild banana (*Musa acuminata*), first reported by Simmonds and Shepherd (1952). Walsh identified wild banana (*mgombatumbili* in KiSwahili) among famine plant foods on Pemba in the modern era (Walsh 2009).

Fauna. The only large mammals present in the Pemba region other than sea mammals (dugong, dolphin, whale) are humans and the domestic animals they introduced (cattle, donkeys, goats, pigs). Non-human primate species include two cercopithecids (vervet/ green monkey and Zanzibar

Table 3.1. Mangrove taxa indigenous to eastern Africa (from Dharani 2002: 299-306; Koenders 1992b)

Taxon	Common name / *Swahili name*	Uses
Combretaceae *Lumnitzera racemosa*	*Mkandaa-mwitu, mkandaa-dume*	Structural poles, fuel
Meliaceae *Xylocarpus granatus*	*Mkomafi, mtonga*	Construction timber, fuel, fruits (food, medicine)
X. moluccensis	*Mkomafi*	Boat construction, fuel
Rhizoporaceae *Ceriops tagal*	Yellow mangrove /*mkandaa*	Structural poles, dye, fuel
Rhizophora mucronata	Red mangrove / *mkoko, mkaka*	Structural poles (exported), fuel, basketry, fish-traps
Sonneratiaceae *Sonneratia alba*	*Mlilana, mpia, mpira*	Boat timber, fish-net floats, low quality fuel, camel fodder
Sterculiaceae *Heritiera littoralis*	*Mkungu, msikundazi*	Fuel, construction timber
Verbenaceae *Avicennia marina*	White mangrove/ *mchu, mtu*	Timber, poles (not structural), fodder, dye, perfume, medicine

[3] Common are members of the following families and genera (italicized): Bignoniaceae, *Bombax*, Chrysobalanaceae, Combretaceae (esp. *Terminalia)*, *Croton*, *Erythroxylum*, *Barringtonia*, *Pterocarpus* , Meliaceae, Moraceae, *Olea*, *Pandanus*, *Pittosporum*, *Quassia*.

[4] For example, members of the Apocynaceae, Connaraceae, Menispermaceae, and Passifloraceae families, as well as *Grewia* and, *Cissus*.

red colobus) and one galagine (greater bush baby). Other medium-sized mammals include: the Pemba blue duiker, large fruit bats including the endemic Pemba "flying fox", three species of mongoose, and an introduced species of civet (Koenders 1992). Jaguars have been sighted on Zanzibar island, but no such reports exist for Pemba (Goldman and Walsh 2002). Pembans, unlike other Swahili-speaking peoples, are reported to hunt flying foxes (*Pterocarpus voeltzkowi)* as a source of tasty meat, although this practice has waned due to the diminishing natural habitat of this large bat (Walsh 1995). Thus while plant and marine foraging could have readily supplemented fishing, farming, and bovid pastoralism on Pemba, hunting wild game was likely not as prevalent on Pemba as it could have been elsewhere along the Swahili coast. This is attested to by zooarchaeological work conducted on several Pemban sites, in which non-domesticated mammal remains are greatly outnumbered by bovid remains (Cain in LaViolette et al. 2003; Mwebi 2000 in Fleisher 2003).

Pemba Island

Pemba is a 68 km x 19 km coral sea mount located 65 km off the shore of Tanzania, separated from the mainland by the deep (700 m) Pemba Channel. Unlike the shallow coastal shelves separating Zanzibar and Mafia islands from the mainland, the depth of the Pemba Channel likely represents a graben, which is a rift in the continental shelf (Schlüter 1997:245). Named the "Green Island" by visiting Arabs, Pemba supports lush vegetation, particularly on its western half, and its high agricultural productivity today suggests a plentiful food supply in the past (Pritchard 1975:162).

The islands of Pemba and Zanzibar receive between 1250 and 2000 mm of rain annually (Pritchard 1975:56) which fall unevenly throughout the calendar year. The "long" rain (*misika*) season starts in March and lasts until the end of May; this is the time when many crops are sown, including rice, cassava, and maize. The "short" rainy season *(vuli)* tends to bring severe rains in late November and December, and may be preceded by an abbreviated rainy season in September. Both Pemba and Zanzibar have bilateral vegetation distribution based on the quality of soils. The eastern halves of both islands bear thin soil exposing coral patches, and receive less rainfall, resulting in poorer agricultural lands. The western regions, however, bear thicker soils and enjoy higher rainfall, which makes them more desirable places for cultivation.

Pemba's and Zanzibar's fertility was exploited duringOmani and British colonial rule in the 19th and early 20th centuries to cultivate lucrative cash crops, most notably cloves. During his surveys of Pemba in 1911-12, Craster (1913) noted that a full one quarter of Pemba was devoted to clove cultivation, and described difficulty while conducting a survey using theodolites due to the dense, hilly vegetation on the western half of the island (see Figure 3.7). Clove cultivation still persists on Pemba today in small-scale fashion, but the large plantations of the past declined after the abolition of slavery and the over-saturation of the global clove market (Fitzgerald 1970[1898]).

Swahili Settlements and Lifeways

Swahili Origins and Culture History

There has been some debate over whether the Swahili originally derived from Bantu agricultural stock (Nurse and Spear 1985) or from a pastoralist tradition (Allen 1993). The Swahili people most likely originated early in the first millennium A.D. from a group of farmers in the northern region of the coast who spoke a Northeast

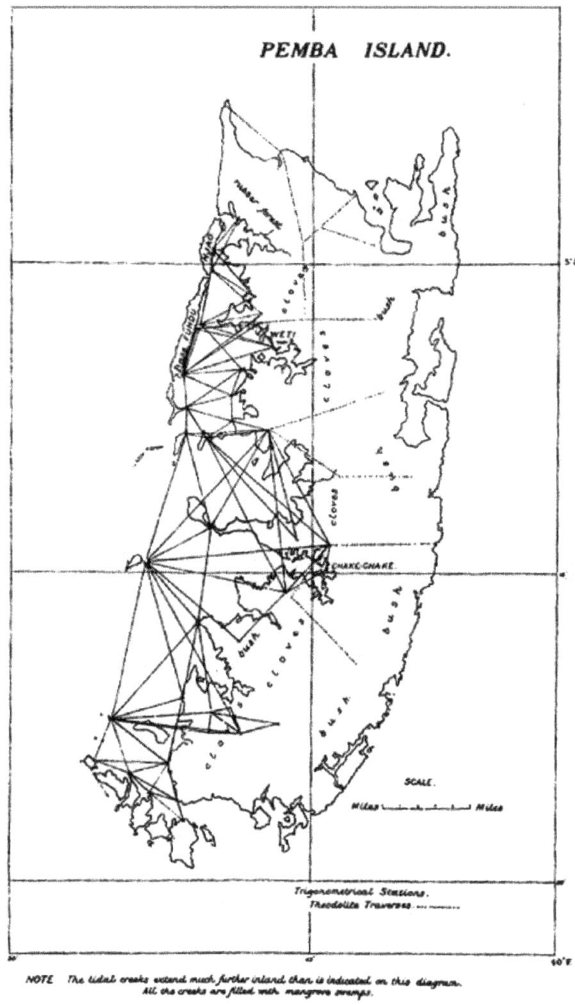

Figure 3.7. Distribution of clove plantations on Pemba Island in 1912. From Craster 1913.

Coast Bantu language of the Sabaki group and who made iron tools and Kwale pottery (EIW in Chami 1994, 1998); (Abungu 1994-95; Horton and Middleton 2000:27; Kusimba 1999:34). At this time, the Swahili are thought to have lived in small villages, relying on fishing, pastoralism, and cereal farming for subsistence (Horton and Mudida 1993;

Wright 1993). Our knowledge of these early villages is extremely limited; better known are the Swahili communities that later blossomed into "stonetowns", so called because they contain stone built structures including mosques, tombs, multi-story houses, and, more rarely, town walls (Fleisher 2003:2). These stonetown sites have received the majority of intellectual investment in the Swahili, due to their prominence on the geographic and historical landscapes, which will be discussed and theorized alongside the discussion of Swahili urbanism below. Several of the stonetown sites excavated on the Swahili coast display similar growth patterns (Kilwa, Manda, Shanga, Pate; see Figure 3.8). They began as small villages in the eighth to tenth centuries growing in economic and political importance between the eleventh and thirteenth centuries, and culminating in large centers by the fourteenth to sixteenth centuries. For example, the site of Manda grew to 10 ha in size by the thirteenth century, with distinct areas of stone masonry appearing separate from neighborhoods of earth and mud structures (Chittick 1984). Kilwa expanded from a 1 ha village of earthen structures in the ninth century to a 30 ha settlement in the eleventh century, with coral block masonry used in its center (Chittick 1974). This change in settlement pattern is accompanied by other changes in Swahili lifeways. Although local pottery is present in all time periods, foreign wares become increasingly important through time (for a thorough treatment of the varieties of ceramics found at Swahili sites, see Horton 1996 or Fleisher 2003:Ch 6). Pottery from South and Southwest Asia and China are present in small quantities early in the sequence (Rouguelle 1996), but comprise an increasingly large portion of the pottery assemblage through time (Chittick 1974, 1984; Horton 1996). Archaeologists and historians also report an increase in Indian textiles and beads through time (Kusimba 1999:131). Between A.D. 1300 and Portuguese encounters with Swahili cities by 1500, stonetowns bear increasing evidence of wealth.

Swahili towns also show an increase in Islamic influence through time. Prior to the second millennium A.D., very few burials appear to be Islamic (Kusimba 1999:109-112), and structures that could be identified as mosques are present only at Shanga (Horton 1996). The larger towns on the Swahili coast embraced Islam by the thirteenth century A.D., exemplified by small mosques built of stone at Kilwa (Chittick 1974:59-99), Shanga (Horton 1996), and Ras Mkumbuu on Pemba Island (Kirkman 1966), and by Islamic influence seen at several cemeteries (Wright 1993). Large-scale acceptance of Islam is evident by A.D. 1000 at Shanga ((Horton 1996: 401), after which time religion was an integrating force among communities along the Swahili coast and a glue in Swahili social fabric (Wright 1993).

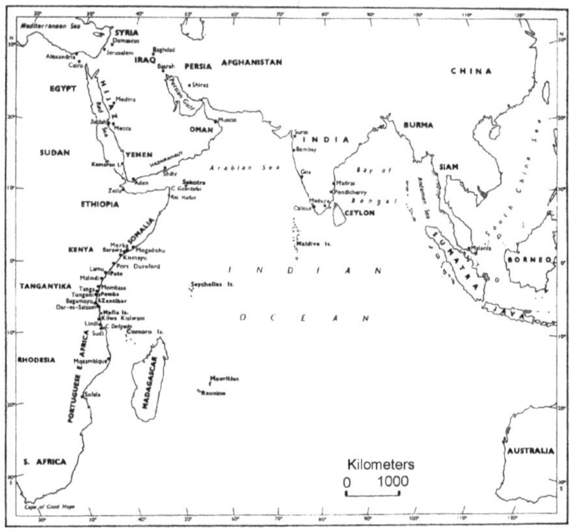

Figure 3.8. Indian Ocean rim showing major ancient and modern cities. Adapted from Freeman-Grenville 1962 Map 1.

Swahili Subsistence Economy

Food production appears to undergo an intensification of domesticated resources during Swahili urbanization, but conclusions are tentative due to scant archaeological evidence. Cereal cultivation was probably important by the tenth century A.D., but prior to the present study flotation was conducted only at several sites from the Comoros Islands (Wright 1984, 1992). Hoffman (in Wright 1984) reports several plant taxa from ninth and tenth century contexts, including Asian rice, millet (possibly *Setaria*), unidentified grass seeds, coconut fragments, and, less frequently, goosefoot (*Chenopodium*) and representatives of the citrus and legume families. Johnson (in Wright 1992) identified small quantities of Asian rice and coconut in thirteenth century deposits, showing a continuation of rice and coconut cultivation practice on the Comoros. A sorghum concentration shows direct evidence for cereal use, and probably storage, at Kilwa (Chittick 1974:52), but cereal agriculture elsewhere is inferred from the presence of grinding stones and the testimonies of early travelers to the coast. The stone terracing indicative of agricultural intensification found at a few locales in eastern Africa (Soper 1997; Sutton 1986) is absent at Swahili sites.

There is at present little evidence to support a pastoralist economy among the early Swahili communities from which faunal evidence has been obtained. Local marine resources may have been more important than domesticates in the earliest Swahili communities. Faunal remains at Pate by the tenth century A.D. suggest a reliance upon fish, with some remains of goat, cattle, and chicken (Wilson and Omar 1997). At Shanga, chicken, sea turtle, dugong, and shellfish appear important at the earliest levels (late 8th century A.D.), followed by an increase in both fish and domesticated mammals (cattle, sheep/goat) through time until approximately A.D. 1000 when domesticates overtake fish in importance (Horton and Mudida 1993; Mudida 1996). Subsistence patterns indicate an intensification of domesticated animal resources by the thirteenth century: there is an increased reliance upon cattle and chickens in general (Wright 1993). Of particular interest is the spatial distribution of fauna at Shanga, where some of the community appears to have had increased access to cattle and camel, while another sector shows higher frequencies of chicken and ovi-caprids, as well as antelope, rat, and dog (Mudida 1996:392-393). This suggests heterogeneity in the Shanga population, where some economic or socio-political segments of the community may have had differential access to food resources.

On Pemba Island, comparisons between stonetown and village animal food consumption show more similarities than differences between A.D. 1000 and 1500 (Cain 2005; Mwebi 2000 in Fleisher 2003). Both village and town contexts show evidence of a dominant maritime food economy across all time periods, including shell taxa, shallow water fishes, and occasionally deep water fish and mammals; shell frequency declines after A.D. 1300, however (Msemwa 2000 in Fleisher 2003). Domesticated cattle dominate the terrestrial fauna, and few goats or wild land mammals appear archaeologically at the four sites sampled, although faunal preservation conditions were admittedly poor at all sites except Chwaka (Cain 2005). Chwaka stonetown contexts show a slightly higher frequency of cattle and chicken remains, and less diversity within the fish taxa represented, compared to the contemporaneous village context (Kaliwa). These zooarchaeological data from northern Pemba will be revisited in more detail in Chapter 8 when considering the full complement of food production patterns on Pemba Island.

Swahili Urbanism: an Identity Crisis

Excavations to date show that Swahili stonetown settlements along the coast shared many features of material culture, architecture, religion, and economics. But the coast was never integrated politically into one entity until Omani colonization in the nineteenth century, which leaves the discipline open to multiple definitions of and working hypotheses about the unit of analysis in Swahili studies. Archaeologists and historians working in the region have variously addressed Swahili "states" (Kusimba 1999), "city-states" (Abungu 1998; Kusimba 1997; Sinclair and Håkansson 2000), "cities" (Chittick 1974; Kirkman 1954), and "ports" (Chittick 1984; Datoo 1970; Stiles 1992), with many scholars today addressing politically neutral categories such as Swahili "communities", "settlements", and "society", or the coast in its entirety. Wilson (1982) proposed five classes of Swahili settlement based primarily upon size: solitary structures; hamlets (a few structures gathered around a mosque); medium sized sites between 2.5 and 5 hectares; larger sites under 15 ha; and larger sites greater than 15 ha. The latter two categories encompass what are commonly understood as "stonetowns", defined as urban centers with distinctive coral and stone monuments that were important facilitators of the long distance trade network and centers of Islamic teaching and practice (e.g., Chittick 1974, 1984; Horton 1996).

Early considerations of the Swahili coast were dominated by an Arab-centric perspective, in which Swahili towns were seen as the direct result of Arab intervention and inspiration, and the Swahili people themselves the legacy of intermarriage between Arab merchants and local African women (Garlake 1966; Kirkman 1966). This Arab-dominant model shaped the archaeological and historic records by limiting attention to stonetowns, which were interesting precisely because they were deemed foreign, as opposed to the "African" villages in the surrounding countryside, which were under-investigated by early researchers (see review in LaViolette and Fleisher 2005). The attribution of Swahili centers to foreign powers was not unique, but rather followed the trend of European colonial thought toward Africa and other colonies in the later nineteenth and early to mid twentieth centuries (Gilbert and Reynolds 2004; Maxon 1994). Cities were considered a hallmark of civilization; therefore, sub-Saharan African cities were evidence that civilizations of foreign origin had once been present in Africa, something worth investigating by colonial governments in their mandate as a civilizing influence (Robertshaw 1990). Since achieving independence, African countries have turned to their rich cultural heritage as a source of pride, to build state cohesion through nationalism. The best example is perhaps Zimbabwe, which is the Shona word for an important structure belonging to a chief, including a court, house, or grave, the best known example of which is the archaeological site of Great Zimbabwe (Kuklick 1991).

Research into Swahili language, culture, and origins has revealed that Swahili society is an African phenomenon (Middleton 1992; Nurse and Spear 1985), and through archaeological evidence stonetowns have been reinterpreted as largely local developments (Horton 1996). Three models have been forwarded in recent years that seek to explain the rise of urbanism in the form of stonetown centers along the Swahili coast. All three share an emphasis on the stonetown as the functional center of an area, controlling the economic, political, and/or ritual foundations of Swahili society. Kusimba (1999) considers stonetowns to have arisen as administrative centers of city-states. In this model, each stonetown governs a hierarchy of settlements in the production of goods destined for long-distance trade. Conversely, Horton and Middleton (Horton 1996; Horton and Middleton 2000) have highlighted the autonomy of Swahili stonetowns, and support a model of relatively independent mercantile centers that arose in order to facilitate production of long distance trade goods and to house marketplaces. Wright (1993) views stonetown development in a more political than economic light. He has developed a model whereby urban centers arose in response to competition among elite families along the coast, who displayed wealth through the accumulation of exotic trade goods, and cemented community support and respect through Islamic practice (Wright 1993).

Much of our current understanding of Swahili urbanization rests upon archaeological investigations of stonetowns, particularly the stone monuments and houses within stonetowns. Coral and stone construction, however, was predominantly used in monumental architecture (mosques, tombs) and in houses of the Swahili elite (Donley-Reid 1990), leaving significant portions of stonetowns and rural villages unexplored. Many questions remain regarding the role of the Swahili majority, those who lived in earth-and-thatch houses in villages and stonetowns. As a result, commoners are conspicuously absent in models of Swahili urbanization, only implicitly drawn as a relatively homogenous countryside performing production tasks required for stonetown economic success. Recent scholarship has focused attention on the interactions between coastal cities and the interior hinterlands to gain a more balanced view of the origin, nature, and function of stonetowns in regional perspective (e.g., Abungu and Mutoro 1993; Haaland 1994-95; Kusimba and Kusimba 2000; LaViolette et al. 1989; Mutoro 1998). But considering the stonetown's dependence upon rural production and ritual support, emphasized in the aforementioned functional models of Swahili urbanization, it is crucial to investigate the extent to which urban and rural commoners were involved in the local and global economies supporting emerging stonetowns if we are adequately to test these models (LaViolette and Fleisher 2005). We now turn our attention to one such investigation of urban-rural connections on the island of Pemba, off the coast of Tanzania.

The Pemba Archaeological Project: New Perspectives on Urban-Rural Dynamics

Archaeological investigations on Pemba Island, Tanzania, by LaViolette, Fleisher, and Mapunda sought to determine the social and economic interactions between a Swahili stonetown (Chwaka) and its immediate rural hinterland (LaViolette, Fleisher, & Mapunda 2003, 2004). This project was based upon intensive excavations at Chwaka, a historically documented Swahili stonetown (Buchanan 1931; Kirkman 1966), as well as surrounding earth and thatch villages detected through recent surveys on northern Pemba by Fleisher (2001, 2003; Fleisher and LaViolette 1999b). Fleisher and LaViolette (2013) have reported the results of archaeological investigations at the 7th-10th C town of Tumbe, a town located adjacent to Chwaka but whose occupation preceded that of Chwaka, with apparently no cultural continuity between the two sites.

While studies of Swahili urbanization have traditionally documented *consumption patterns* within stonetowns, particularly of imported goods, a new and valuable perspective is achieved through the documentation of *production patterns* in both stonetown and village communities (Brumfiel and Earle 1987; Costin 1991; Wattenmaker 1998:3). Examples of local Swahili production include industries involved in making iron products, ceramics, beads, textiles, and plant foods. The Pemba Archaeological Project focuses attention on the house as a unit of analysis in understanding patterns of material consumption and production.

Swahili Settlement History on Pemba Island

The earliest evidence for the settlement of Pemba comes in the form of numerous small settlements during the eighth to tenth centuries A.D. (Fleisher 2003:137-139). By A.D. 1100, larger villages and towns on Pemba's coast had emerged that were surrounded by considerable communities, which extended well into the island's interior. Between A.D. 1100 and 1300 people aggregated into larger stonetown settlements, located predominantly on the coast, and abandoned the majority of nearby villages. Thus, stonetown expansion is seen by Fleisher (2003:139) as a net movement of the rural population into larger centers, and not an overall increase in Swahili population as argued by Kusimba (1999) as being important to stonetown development. Several of these large centers were, in turn, abandoned during the fifteenth and sixteenth centuries, including Chwaka. This pattern suggests that stonetown centers may have wielded considerable influence on Pemba, particularly in their ability to attract or command population migration from the countryside into the city.

Both historic accounts and archaeological data provide evidence that Pembans actively participated in the Indian Ocean trade network, and that some individuals accumulated great wealth (Kirkman 1966; LaViolette 2000). Pemba supported high populations, attested to by the concentration of both stone works (Garlake 1966; Kirkman 1966) and recently discovered sub-surface archaeological sites (Fleisher 2001, 2003). Historic accounts cite Pemba as supporting important centers before the arrival of the Portuguese (Buchanan 1931; Pearce 1967 [1920]). Early Arab writings reveal that Pemba had a partially Muslim population in the tenth century, housing some of the earliest Muslim towns on the coast (Chittick 1971:112). Prior to Portuguese contact at the close of the fifteenth century, the socio-political organization of Pemban communities was characterized by central control of a stonetown by a sole ruler who inherited the position (Horton and Middleton 2000). This system of rule, similar to that of other coastal sites south of Mombasa, known as the Shirazi mode of dominance, contrasts with the republic-style Waungwana mode found on the northern coast, which appears to be based on consensus of elders (Allen 1993). This characterization belies a diversity of political forms found throughout the Swahili world, however, particularly systems which occupy a place on the continuum between Shirazi and Waungwana modes.

Project Excavations

Building upon the previous investigations at the known historic ruins of Chwaka (Buchanan 1931; Kirkman 1966; LaViolette 2000) and sub-surface sites located through extensive surveying of northern Pemba (Fleisher 2001, 2003; Fleisher and LaViolette 1999b; LaViolette and Fleisher 1995), excavations between 2002 and 2006 by LaViolette, Fleisher, and Mapunda have targeted four sites (see Figure 3.9). Both Tumbe and Chwaka are extensive (>10 ha) settlements located on a ridge overlooking Micheweni Bay that is close to cultivable lands (see Figures 3.10-3.11). Tumbe was first occupied between the seventh and tenth centuries, and was abandoned less than a century before the adjacent site of Chwaka was founded around A.D. 1050. Mosques and tombs of coral rag architecture erected at Chwaka, many of which are visible above-ground today, attest to a stonetown presence here until the late fifteenth or early sixteenth century. Kimimba and Kaliwa are smaller village settlements bearing earthen structures. Kimimba is temporally associated with Tumbe and located 2 km distant, thus the people in this town and village likely interacted and their economies may have intersected. The archaeobotanical findings from Tumbe and Kimimba are presented together in Chapter 5. Kaliwa is located 2 km northwest of Chwaka and dates to the later half of Chwaka's occupation. The floral finds from Kaliwa and Chwaka are presented in tandem in Chapter 6.

Much of the information on the four sites given below comes from numerous publications authored by site directors Adria LaViolette, Jeffrey Fleisher, and Bertram Mupunda since 2002 including peer-reviewed articles (Fleisher and LaViolette 2013; Flexner *et al.* 2008; LaViolette and Fleisher 2009) and project reports (LaViolette *et al.* 2003, 2004). Some of these also contain data derived from a 2006 excavation season of contexts not included in this archaeobotanical research.

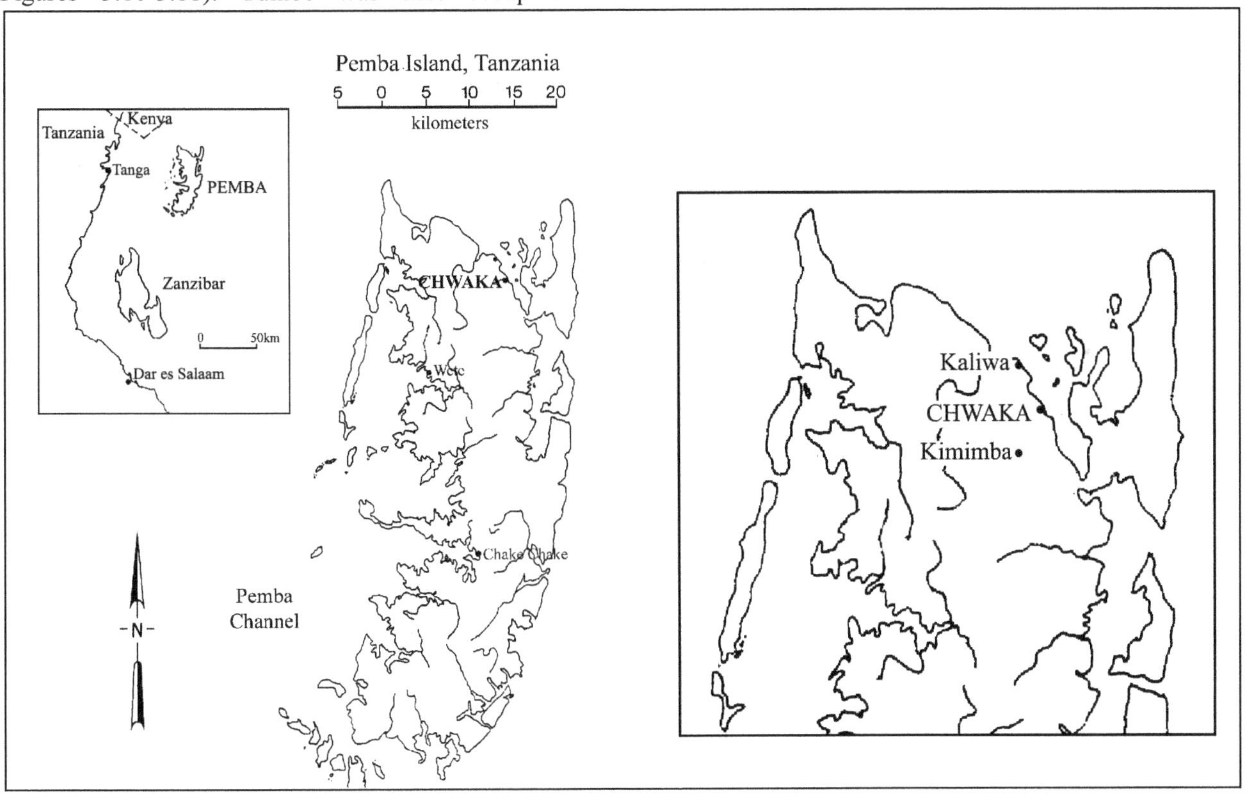

Figure 3.9. Maps of Pemba Island, Tanzania, showing the location of major modern towns (Tanga, Dar es Salaam, Chake Chake, Wete) and archaeological sites featured in this study (Chwaka/Tumbe, Kaliwa, Kimimba). Adapted from Fleisher 2003:143 Figure 4.1, ibid:191 Figure 5.1.

Based upon relative dating through analysis of imported pottery, confirmed by radiocarbon dates on wood and AMS dates on charred plant remains (described later), I use the following periodization:

Period 1: 600-1000 A.D.
Period 2: 1000-1100 A.D.
Period 3: 1100-1300 A.D.
Period 4: 1300-1500 A.D.
Period 5: 1500-1700 A.D.
Period 6: 1700-1850 A.D.

Tumbe (Seventh to Tenth Century Town). The site of Tumbe contains evidence for two episodes of occupation: the earliest is termed the Tana level due to the signature local pottery tradition, and dates to Period 1. Tumbe's Tana levels are relatively artifactually dense, considering its lack of stone architecture, and contain a variety of artifact classes, including: local pottery, three types of imported pottery (Sasanian-Islamic, white-glaze ware, and Siraf storage jars), glass vessels, beads, copper pieces, iron slag and tuyeres, daub, coral, unmodified bone, shell, and pottery sherds with incised linear grooves considered to be "bead grinders" (Fleisher and LaViolette 2013; Flexner *et al.* 2008). Tumbe also contains a significant amount of high status goods – for example, the imported sherd ratio (number of imports for every 100 local sherds) for Tumbe is 3.5, which is comparable to that of Shanga (3.6), an important stonetown in the Lamu archipelago, during the same time period. Both local and imported pottery styles from Period 1 are medium to large rounded-jar shapes, most likely for transporting fluids in the case of the imports, and for cooking, storing, and serving semi-fluids in the case of local wares. All structures located to date are composed of earthen material.

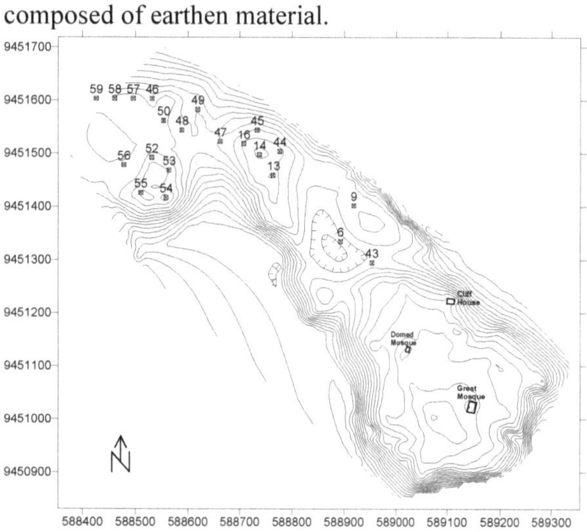

Fig 3.10. Distribution of test excavation units bearing Period 1 (A.D. 600-1000) artifacts at Tumbe. Note proximity to Chwaka mosque structures (see figure 3.11 for location of Chwaka units). From LaViolette et al. (2003:24, Figure 1).

Mazrui levels overlie Tana levels, often with an interceding stratum of mixed or sterile sediment. Mazrui deposits date to the eighteenth and nineteenth centuries and represent a migration of mainland coastal families of Omani origin to the island. In terms of artifacts, Period 5 levels are easily distinguished from earlier occupation levels. Archaeobotanically, however, these distinctions are more difficult to make, particularly because intermediary sediments contain mixed artifact assemblages. Thus only the lowest level or two within each Tumbe unit can reliably be associated with Period 1 (see Figure 3.10); this generally corresponds to layers three or four, and occasionally five. Direct radiocarbon dating of individual plant remains are reported in Chapter 4.

Kimimba (Eighth to Tenth Century Village). Kimimba was discovered during shovel-test pits excavated by Fleisher in 1999 (2003: 182-186) in the absence of any surface deposits. There is no evidence of further settlement after its abandonment, and disturbance from agricultural activities appears to be shallow. All Kimimba contexts with cultural materials are considered to date from the same time period, based upon the relative chronologies of local and imported wares. Excavations in 2002 uncovered evidence for several structures; many post-mold features were discovered, as well as daub fragments and one intact charred wooden post. Local pottery clusters to form two areas of concentration throughout the site and positively correlates with the distribution of architectural remains. This suggests the use of abandoned houses as refuse areas (LaViolette et al. 2003). Imported materials occur at low density and appear to follow a different density distribution than that observed for local pottery and architectural features. This may represent socio-economic differences among households at Kimimba (LaViolette et al. 2003). Kimimba's imported sherd ratio (number of imports per 100 local sherds) has a particularly low value (0.2) when compared to those at Tumbe (3.6) and also at Shanga (3.5), a coastal site of the same time period.

Chwaka (Eleventh to Sixteenth Century Stonetown). Chwaka has long been known as an archaeological site due to the presence of standing stone ruins, including two mosques, two houses, and ten tombs. Chwaka is located approximately 100 m to the southeast of Tumbe, but at present there is no evidence for overlapping occupation of these two sites (see Figure 3.11). A radiocarbon date run on Chwaka's earliest layer yielded a calibrated date of A.D. 1026 (± 30), while the occupation of Period 1 Tumbe deposits ends by A.D. 950 (Fleisher and LaViolette 2013; LaViolette *et al.* 2003). Chwaka's location at the end of the peninsula seems to be strategic in terms of access to the ocean, but it is raised up sufficiently to be defensive as well. This stonetown settlement was likely an important port on the Pemban coast based on its size, location, and ruins, but archaeologists know little about the local economies that supported participation in the Indian Ocean trade network, at

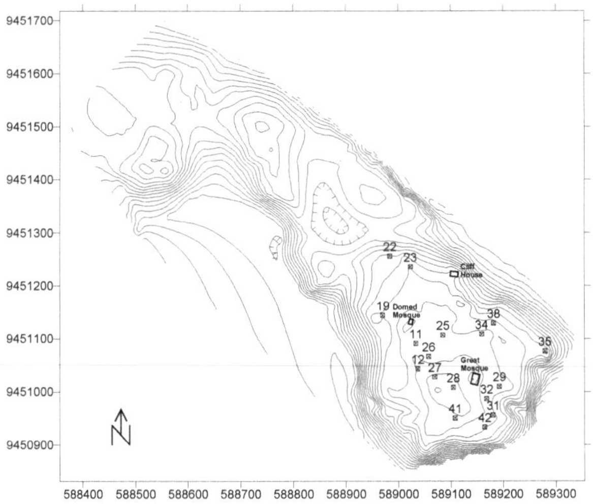

Figure 3.11. Distribution of test excavation units bearing Period 3 (AD 1100-1300) artifacts at Chwaka. From LaViolette et al. (2003:26, Figure 3).

Chwaka or elsewhere along the coast. This is largely because excavations have traditionally focused upon stone architectural components. In contrast, work by Fleisher and LaViolette, including that reported here, has sought to include earthen architectural features and their material remains to illuminate the lives of the Swahili majority (Fleisher and LaViolette 1999a,b).

Like Tumbe, Chwaka contains a high density and diversity of artifacts. The locally-made pottery appears to consist largely of open vessels, such as bowls and plates, as opposed to the jar-rich local pottery assemblage at Tumbe (LaViolette et al. 2003). This suggests that perhaps Chwakans were using vessels for sharing and displaying foods (Fleisher 2003), a theory that I will return to in Chapter 9.

Interestingly, there is a decrease in the imported-sherd ratio through time at Chwaka: the combined average is 1.2, which is lower than contemporaneous stonetown settlements elsewhere along the coast (LaViolette *et al.* 2003). The lower import ratio could be due to the increased sampling of non-elite contexts at Chwaka, as opposed to excavations of comparable sites where focus was placed on wealthy stonetown contexts. Imported pottery is not evenly distributed throughout the site, with higher proportions recovered in some units suggesting differentiation in access to wealth among the residents of Chwaka. Thus, the low average ratio for Chwaka's imports may in fact be explained by the inclusion of non-wealthy, and therefore non-importing, households into the average (LaViolette *et al.* 2003). Other artifact classes include glass vessels, beads (glass and carnelian), metal fragments (Fe, Cu, Pb), tuyeres, sherds modified into "bead grinders", spindle whorls, ground sherds, groundstone pieces, lamp fragments, daub, unmodified bone, shell, and coral. This variety holds with findings from similarly dated coastal sites. Surprisingly, the material remains of iron-working (mostly slag, but also tuyeres, which are ceramic pipes used to draw air into furnaces) are present in virtually every unit. Moreover, both spindle whorls and bead grinders are found throughout Chwaka, suggesting widespread production activities. In Chapter 6 I discuss the evidence for another arena of Chwaka production, that of food production.

Kaliwa (Fourteenth to Fifteenth Century Village). Kaliwa was found during Fleisher's 1999 surveys, and was excavated in 2004 with the assistance of the University of Dar-es-Salaam's Archaeology Unit field school (Fleisher 2001, 2003; LaViolette et al. 2004). The site is located close to the boundary of the modern town of Tumbe near the present-day fishing beach and market, a few kilometers northwest of the archaeological sites of Tumbe and Chwaka. Investigations here have revealed an earthen house floor in one context supporting three iron-furnace features. The site also shows rich midden stratigraphy. Surveys conducted in 2004 along the nearby beach produced additional evidence for iron-working, including one tuyere pipe (ceramic furnace hole for bellows) encased in melted iron and large quantities of iron slag. Further work is planned, including analysis of imports and submission of carbon samples to secure dates for this site.

Swahili Plant Use and Food Production

Archaeobotany in Eastern Africa

Archaeologists and paleoethnobotanists are only beginning to contribute to discussions of ancient plant use in eastern Africa. Indirect evidence for plant use includes the presence of agricultural tools, grinding stones, storage structures, irrigation works, and grain impressions in pottery (Haaland 1994-95; Soper 1997). Direct evidence for plant use, in the form of macrobotanical remains recovered from archaeological sites, is rare in eastern Africa despite commendable attempts at recovery (Ambrose 1984; Kabuye 1991; Marshall and Hildebrand 2002; Neumann 2003; Reid and Young 2000; Robertshaw and Wetterstrom 1989; Sutton 1987; Wetterstrom 1991; Young and Thompson 1999; see recent success through systematic flotation by Alison Crowther in Helm *et al.* 2012.)

Low macrobotanical recovery rates are attributed to poor preservation of identifiable plant remains (Robertshaw and Wetterstrom 1989; Sutton 1987; Wetterstrom 1991) due to a combination of sparse deposition and rapid destruction by organisms within tropical soils. Another factor is that the preparation and discard methods used in eastern African cuisine, such as brewing beer or making porridge or stew, may have precluded the charring or deposition of plant remains (Young and Thompson 1999). In at least one instance, the lack of carbonized plant remains resulted from the use of inappropriate recovery methods (Kyule 1991).

Archaeobotanical studies in northeastern Africa have yielded unparalleled finds that are instructive in considering the plant foods available in eastern Africa through maritime trade. At Roman trade ports (Quseir al-Qadim, Berenike) on the Red Sea coast of Egypt, desiccated (and, to a lesser extent, carbonized) plant remains are found in great abundance signaling trade between the Mediterranean, Asia, and Africa (Cappers 2003; van der Veen 2003a; Wendrich et al. 2003). African taxa found in Roman (first to sixth centuries A.D.) contexts in Egypt include: baobab (*Adansonia digitata*), tamarind (*Tamarindus indicus*) and Ethiopian pea (*Pisum abyssinicum*) (see Appendix A for plant names and authories). The baobab seed is of notable interest as it likely originated in sub-Saharan Africa and may be direct evidence of connections with "Rhapta", a trade destination on the eastern coast of Africa known from texts but not yet identified archaeologically. Plants of south or southeast Asian origin found in Roman Egypt include black pepper (*Piper nigrum*), mung bean (*Vigna radiata*), Asian rice, coconut (*Cocos nucifera*), *Phyllanthus emblica*, and Job's tear (*Coix lacryma-jobi*) (Cappers 2003). At Aksum, significant quantities of plant remains were found in pre-Aksumite (circa 500 B.C.) and Aksumite (A.D. 500-700) deposits, where crops of both Near Eastern (wheat, barley, flax, lentil) and African origin (tef, finger millet, sorghum, noog, cotton) were identified (Boardman 1999; 2000). Further south, in present-day Zimbabwe, the first domesticates appear in the first millennium A.D., including sorghum, finger millet, cowpea, bottle gourd, and watermelon (Jonsson 1998:108). Most recently, scholars associated with the Sealinks project have conducted archaeobotanical investigations in coastal Kenya, Tanzania, the Comoros and Madagascar to determine how Indian Ocean communities facilitated the movement of plants approximately 1000 BCE to 1000 CE. Preliminary results from this project have reported African staples sorghum, pearl millet, finger millet and baobab from two Early Iron Age sites in Kenya (Helm *et al.,* 2012).

Reconstructing Swahili Plant Use

Precisely which crops were cultivated by the Swahili is a subject for speculation because there is only scant archaeobotanical evidence to demonstrate where and when individual crops first appear in eastern Africa (Blench 1994; Harlan 1993; Harlan et al. 1976; Neumann 2003; Wetterstrom 1991). Likely candidates include the traditional African crops: sorghum, pearl millet, finger millet, cotton, cowpea, hyacinth bean, and yam (Harlan 1993, 1997; Harlan et al. 1976; Marshall 1998; NRC 1996; Portères 1976; Wetterstrom 1991). Taxa domesticated originally in Asia are hypothesized to have arrived in Eastern Africa as early as the tenth century A.D., including: banana (*Musa paradisiaca*), taro (*Colocasia esculenta*), coconut, rice, sugar cane (*Saccharum officinarum*), and the date palm (*Phoenix dactylifera*) (Stahl 1984; Watson 1983). Additionally, there is an argument for the prehistoric introduction of the sweet potato (*Ipomoea batatas*) from South America via Oceania (Blench 1994) . Crops of New World origin that form a significant portion of the diet today were probably introduced by the Portuguese (sixteenth and seventeenth centuries) or during German and British colonial rule (mid-nineteenth to mid-twentieth centuries). This category includes cassava (*Manihot esculenta*), potato (*Solanum tuberosum*), tomato (*Lycopersicon esculentum*), and maize (*Zea mays*). Pembans today grow rice and cassava in large fields, and use gardens or field edges to grow coconut, banana, taro, maize, sorghum, tomatoes, potatoes, sweet potatoes, eggplant, legumes, and numerous fruit tree taxa (Koenders 1992; Williams 1949).

Macrobotanical Evidence. As discussed previously in the context of Swahili dietary patterns, plant remains were known only from a few Swahili sites prior to the present study. These include Kilwa, where a concentration of sorghum grains was recovered from an ancient house floor (Chittick 1974:52). Flotation conducted on the Comoros Islands has led to the recovery of archaeobotanical remains, although sampling was limited. Dembeni phase (ninth to tenth centuries A.D.) sediments from the site of Sima on the main Comoros Island yielded primarily rice, but also several grains of millet (possibly *Setaria*), four pieces of coconut rind, and possible bean and citrus fruit remains (Hoffman in Wright 1984). Recovery of small amounts of rice and coconut from thirteenth century Sima deposits suggests the continued importance of these crops in the Comoros (Johnson in Wright 1992). Macrobotanical analyses from sites in central Madagascar show that twelfth to fourteenth century sites were dominant in rushes and sedges, while domesticates – rice and cowpea (Vigna unguiculata L.Walp.) – are limited to late fifteenth to eighteenth century sites (Wetterstrom and Wright 2007). Aside from these extremely valuable but limited instances of macrobotanical recovery, evidence of Swahili plant use has largely come from indirect sources, the interpretation of which is fraught with problems both botanical and anthropological in nature.

Evidence from linguistic analysis and historic description. Linguistic and ethno-historic accounts describe the Swahili as farmers, placing human-plant relations at the forefront of their economy and society (Nurse and Spear 1985; Wigboldus 1994-95). Sweet banana (as opposed to plantain) and coconut appear to be important by the end of the first millennium A.D. according to linguistic studies (Blench 1994; Wigboldus 1994-95). In addition to the rice and grain (wheat) documented as imports, coconut may have been known along the coast of eastern Africa according to the second century A.D. author of the Periplus of the Red Sea (see Table 3.2 for summary of historical reports of plants in eastern Africa), although interpretations differ as to the meaning of the word given (Blench 1994;

Huntingford 1980). Travelers to eastern Africa from A.D. 915 to 1800 consistently report the use of rice and millet as subsistence foods. "Millet" is a category of cultivated cereal, possibly referring to sorghum, pearl millet, and/or finger millet, commonly reported by early visitors to the eastern African coast, such as Al-Masudi in A.D. 915 (Freeman-Grenville 1975:16). Portuguese writers noticed reliance upon rice during their 16th century exploration of the Swahili coast, but were most impressed by the abundance and diversity of fruits (Freeman-Grenville 1975:63, 132).

Fortuitously, some historic travelers were particularly enamored with local cuisines, including the famed Moroccan scholar Ibn Battuta. In A.D. 1331 he was served the following dishes by the sultan/sheikh of Mogadishu: "a dish with some betel leaves and areca nuts on it" (Hamdun & Noël 1998:17), and:

> rice cooked with ghee placed on a large wooden dish. They put on top dishes of *kushan*, that is the relish, of chicken and meat and fish and vegetables. They cook banana

Table 3.2. Early reports of plants imported, exported, or found in use along the coast of eastern Africa. Data from (Freeman-Grenville 1962; Hamdun and King 1998:15-25; Huntingford 1980:28-31)

Author/work	Location	Plants/products	Uses
Periplus Maris Erythraea Early 2nd C	"Marts" of Azania	Coconut?, spices	Export
	Brought from Egypt to "marts"	"corn" (wheat or other grain), rice, sesame oil, cloth (cotton?), "honey cane" (sugar cane?), wine	Import
Cosmas Indico-pleustes c. 525	Northern section of east coast	Spices, frankincense, cassia,	Export
Al-Mas'udi c. 915	"Zanj" – east coast	Bananas, millet, coconut, "kalari"	Subsistence
Al-Idrisi (second hand) 12th C	"Zanj" coast and islands	Fruit, sorghum, sugar-cane, camphor trees, bananas, rice,	Subsistence and export (rice)
Chao Ju-Kua (second hand) 1226	"Zanguebar and Berbera coast"	White cotton cloth, red cotton	Import
		Meal, baked cakes	Subsistence
Marco Polo (second hand) c. 1295	East coast	Rice, dates, sugar, spices	Subsistence
Ibn Battuta 1331	Mogadishu	Cloth	Export
		Areca nuts, betel leaves	Chew
		Rice, vegetables, banana, lemon, peppercorns, saffron, ginger, mango	Feasting/ subsistence
	Mombasa	Banana, lemon, orange, *jammun* (rose apple?)	Subsistence
		Grain (rice?)	Import
Vasco de Gama April 1498	Mombasa	Orange, lemon, sugar-cane	Gifts
		Cloves, pepper, "corn"	Exports
	Boat along coast	Maize	Provisions
January 1499	Malindi	Cloves, cumin, nutmeg, ginger, pepper	Gifts
		Palm groves, maize, vegetables	Cultivars
	Malindi	Oranges, fruits	Gifts
Voyage of Pedro Alvares Cabral August 1500	Malindi	Rice, dates, oranges, fruit	Gifts
Voyage of Dom Francisco D'Almeida 1505	Kilwa	Millet, rice, cultivated roots, palms (coconut and other, not date), wild fruits, vegetables, "Guinea maize," betel, oranges, lemons, cotton,	Cultivated/ Foraged/
	Mombasa	Maize, rice	Provisions
		Oranges, pomegranates, lemons, sugar cane	Cultivars
		arrow poisons	Weapons
Duarte Barbosa 1517-18	Sofala, Mombasa, Malindi,Madagscar, offshore islands	Millet, rice, fruits, vegetables	Subsistence
St Francis Xavier 1542	Socotra	Dates only – no grains grown (too arid)	Subsistence
Rezende 1634	Pemba	Rice, "corn", sesame, beans	Rice tribute
Crassons de Medeuil 1784-5	Kilwa	Millet, indigo, cotton, sugar cane	Local products
Dallons 1804	Zanzibar	Rice, maize, millet, coconut	Subsistence

> before it is ripe with fresh milk and they put it on a dish, and they put sour milk in a dish with pickled lemon on it and bunches of pickled chillies, vinegared and salted, and green ginger and mangoes ... When they eat a ball of rice, they eat after it something from these salted and vinegared foods.
>
> (Hamdun & Noël 1998:18)

Ibn Battuta closes his description of dining in Mogadishu with the observation that his hosts ate a large amount of food and were "extremely large and fat of body" (Hamdun & Noël 1998:18). On the island of Mombasa later that same year, Ibn Battuta notes the presence of banana and citrus trees ("lemon" and "citron") and a tree called *jammun* locally. Ibn Battuta also reports that "there is no cultivation of grain among the people of this island: food is brought to them from the *Sawahil* [coast]. The greater part of their food is banana and fish" (Hamdun & Noël 1998:22). This is echoed by Kirkman (1964:179), who notes that "his [the Sultan of Mombasa's] successor Hasan was finally recognized as ruler of Pemba in exchange for an annual tribute of rice to the Captain of Mombasa", suggesting that a rice surplus was generated on Pemba and mobilized for political use. Mombasa was a crucial entrepôt and powerful administrative center of the Swahili world (Middleton 1992:57), and it likely maintained connections with nearby Pemba.

Botanical historiography. Travelers' reports of local plant use can be problematic for several reasons. Firstly, their experiences as guests of lauded officials provide them with a view skewed towards the diets of the privileged, as opposed to the staples of the majority of the population. This viewpoint, however, is advantageous to scholars interested in feasting – the Swahili hosts clearly served high status foods to impress visitors and local attending officials alike. Significantly, Arab visitors report the presence of both millet and rice in local diets along the coast, but attest to being preferentially served rice by their hosts. A second problem in historical reports involves the plant identifications themselves, which can be inaccurate for many reasons, including incorrect reporting, the numerous disjoints in the translation process, and the use of variously interpretable morphological descriptions (Cappers 2003).

Yet another confounding factor is the whim of the writer and the subjects they choose to report. Most ethno-historic accounts of the Swahili report the abundance and variety of fruits grown or served there, because these taxa would have been relatively rare or unknown in the homelands of the authors. Ibn Battuta's description of the rose apple clearly illustrates these issues: "[t]hey have fruit which they call the *jammun*, which is similar to the olive and its stone is like its stone except that it is extremely sweet" (Hamdun & Noël 1998:21). On the basis of morphology, the fruit could perhaps be the jujube (*Zizyphus*, Rhamnaceae), but the local name provides a more promising lead. *Jammun* may be related to the Hindi word for the rose apple and clove genus *Syzygium* (Myrtaceae) (Hamdun & Noël 1998: 80), of which there are five edible fruit species recognized on Pemba (cloves aside) called *mtofaa*, *mzambarau*, and *mgulabi* locally (Koenders 1992). Ibn Battuta's reference to "chillies" in 1331 is also problematic, due to the South American origin of the common chile (*Capsicum* sp.). Similarly, Portuguese writers report that maize was grown on Pemba by 1634 (Freeman-Grenville 1962:182; McCann 2001), but these statements may refer instead to sorghum, which in the field bears some resemblance to maize, and would be called "corn" by some Europeans if associated with grain fed to animals.

Constructing a Model of Swahili Urban Food Production and Consumption

Taken together, the lines of evidence point to rice being the staple of Swahili stonetown residents, likely introduced through Asian maritime connections sometime in the mid-first millennium A.D. Millets were also reportedly grown, and may have been important earlier, perhaps during al-Masudi's time (tenth century A.D.), or in different contexts than those reported by later travelers. Grains were supplemented with a variety of local and introduced fruit, vegetable, and spice plants, as well as fish and cattle and chicken. Taken at face value, the ethnohistoric evidence lends itself better to an interpretation of Swahili elite foodways, and does little to show what the rest of the population ate, apart from passing references to bananas and millet. The ethnohistoric evidence also points to the provisioning of certain elements of the population, or perhaps entire settlements, by the time of Ibn Battuta's visit to Mombasa in 1331. This practice apparently continued and was reported by Portuguese authors.

The archaeobotanical and historic data leave us with several critical gaps in our knowledge, including questions about: 1. the basic subsistence components of the Swahili diet; 2. patterns of food production and distribution among different settlements and classes; and 3. the effects of urbanization on plant food production.

1. **Subsistence -** Much remains to be known of Swahili diet, including the subsistence patterns of the original colonizers of the coast, and how much variation was present among different communities through time, particularly after the rise of urban centers. This can only be addressed through widespread archaeobotanical sampling and analysis. The results presented in Chapters 5 and 6 are a first contribution toward a

regionally specific understanding of Swahili foodways over a millennium of settlement, focusing on northern Pemba.

2. **Food Production** - The reports of some communities, such as Mombasa and Socotra, being provisioned by nearby settlements indicate that some individuals or communities produced an agricultural surplus, and that such a surplus was made available through tribute, tax, or trade. How was this grain surplus processed and redistributed? Did all stonetowns receive food products, or only some? Were foods the product of local exchange, or tax, or tribute?
3. **Urbanization** - As discussed in Chapter 2, and revisited below, urbanization is associated with a centralization and specialization of food resources. If Swahili stonetowns were provisioned by nearby villages, then different archaeobotanical signatures should be evident in these different contexts; namely that villages should bear primarily producer signatures and stonetowns should show mainly consumption activities. Conversely, if stonetown households each produced their own food, then villages and stonetowns should show little difference archaeobotanically. Fleisher (2003) found that northern Pemban stonetowns were connected but not necessarily dependent economically upon one another based on craft production and settlement pattern data. Does food production follow this same pattern?

If Swahili cities were powerful enough to extract large amounts of surplus food products regularly, then this may have created systems of communal labor to process harvested crops for storage in stonetown contexts (summarized in Table 3.3). This process may also have influenced which crops were grown, because demand for a certain food, probably rice in the present case, may have resulted in a decrease in the diversity of plant taxa grown and consumed. Conversely, if food was produced by and stored in individual households, then a different archaeobotanical signature should emerge. Grains would be stored domestically in a less processed form, and taxonomic diversity would remain high. Moreover, every household would show evidence of early crop processing and storage, while communal labor likely produced a surplus to be kept in a few, large, stonetown storage facility. Communal labor would also reduce grains to a form more easily processed after storage, so provisioned households would produce less cleaning waste. Finally, communal and household labor strategies may also be evident by looking at the material remains of trade or tribute relationships bought with grain. If grain was controlled and redistributed through a central figure, office, or clan, and this grain was used to buy trade products or tribute, then the distribution of high-status goods should follow a centralized distribution as well. Conversely, if each household could trade or give the products of its own field, and reap the rewards directly, then high-status goods should show a more even distribution throughout the community.

I hypothesize that the Swahili landscape consisted of towns and villages bonded in units of food production, in which communal production of important subsistence and trade items, likely rice, dominated the local economy. This follows from the consideration that trade, and the local products supporting it, were central to the success of Swahili towns. According to this hypothesis, towns and villages should exhibit different archaeobotanical patterns. Towns would be predominantly grain consumers, engaged in a system of communal production as noted above in Table 3.3. Rural communities would then be the producers of the town's grain supply, and as such show evidence of more production than consumption, likely with domestic patterns of production responsible for consumption patterns within the village itself. Thus, I expect that the towns of Tumbe and Chwaka should show differences from their neighboring villages, Kimimba and Kaliwa, in taxonomic diversity and distribution of grain processing products and waste.

Table 3.3. Expected botanical signatures of communal and household production modes.

Variable	Communal production	Household production
Storage	Central; in stonetown	Domestic, widespread
Pre-storage processing	More, centralized	Less, widespread
Post-storage processing	Less, widespread	More, widespread
Species and part diversity	Lower	Higher
Trade returns/gifts	Concentrated in space	Widespread

Having created a model to test how urban stonetown food surpluses were achieved, I now turn to the archaeobotanical record of four sites in northern Pemba to see how food production changed through time in towns and villages. Chapter 4 presents the methods used in this research, followed by the results, presented in Chapters 5 and 6.

CHAPTER 4. FIELD AND LABORATORY METHODS

Introduction

Three methods were used to collect data about ancient and modern plant use on Pemba Island. Archaeological plant remains were retrieved using bucket flotation, a manual method of water separation. Modern plants were collected as reference specimens following guidelines outlined in the Missouri Botanical Garden's field collection guide (2005) and in standard ethnobotanical texts (Cotton 1996; Martin 1995). Ethnobotanical information was obtained through individual non-structured encounters, semi-structured group interviews, and participant observation of farming practices.

Archaeobotanical Methods – Field

Sampling Strategies

Samples for archaeobotanical analysis were collected in conjunction with ongoing excavations at four sites on Pemba Island, directed by LaViolette, Fleisher, and Mapunda (2003, 2004). Excavations took place from June 17 to July 27, 2002 and from June 16 to July 20, 2004. Sampling and recovery methods were as consistent as possible among the sites and between field seasons to maximize comparability of the results obtained (Wagner 1988). Following the blanket sampling approach, all contexts containing cultural materials were sampled for flotation (Pearsall 2000). Composite samples were taken from each context, meaning that sediment was collected from a wide surface area within the unit (Pearsall 2000:69-71). In 2002, 10-20 liter samples were taken from every arbitrary level within each natural layer at Tumbe, Chwaka, and Kimimba to gain high stratigraphic resolution. In 2004 this rigorous sampling strategy was limited to Kaliwa alone, and 10 liter flotation samples were taken from the top level within each natural layer at Tumbe and Chwaka. In both seasons, opportunistic samples of various sizes (150 ml – 25 liters) were taken from feature contexts, such as suspected hearths and ovens.

Flotation sediment samples were placed in a cleaned woven plastic bag lining a 10 liter bucket using a cleaned trowel. Bags were filled with sediment until the desired level was reached, typically rounding the top of the bucket. Provenience information was written on two pieces of plastic flagging tape; one was put inside the bag, the other was used to tie the bag closed. These pieces of tape and the provenience information they carried followed the flotation sample through to the final analysis. Bags of sediment destined for flotation were brought back to the dig house/laboratory at the end of each field day and stored in the shade until floated, generally within three days.

Flotation Methods

Archaeological sediments were processed using the bucket method, which was chosen because the materials were readily available in the field and because it allowed me to observe the sediments and therefore gauge recovery during processing. Moreover, this system was the most likely to be utilized by future archaeologists working in this region because of its simplicity and low likelihood of cross-sample contamination.

Prior to flotation, the details of each sample and flotation event were recorded in a flotation log. Sample volumes were measured again prior to flotation, and sediment was partitioned into two to four 10 liter buckets for ease of flotation. Water was added to each bucket using a hose, and samples were then agitated by gentle manual action or using water pressure delivered from the nozzle. The water and floating particles were then poured through the light fraction sieve (see Figures 4.1 and 4.2 for flotation sequence), taking care not to pour out the sediment itself. The floating material (light fraction, also called the flot) was caught in a 0.50 mm brass standard geological sieve, although for some test samples the sieve was lined with a fine mosquito netting of approximately 0.35 mm aperture. Water was added to the bucket again, agitated, poured out, and repeated until the water ran clear or no further charred remains could be observed in the sediment. The sediment remaining in the bucket (heavy fraction) was then wet screened. Heavy fractions were poured through a 1.00 mm sieve, or, where large sand grain sizes dominated, a 2.00 mm sieve. Each fraction was put into a mesh bag for drying in the shade over two to three days. When sufficiently dried for storage, each fraction was placed in a labeled plastic bag along with silica desiccant packets for transportation back to the United States. To minimize breakage during transport, bagged light fraction samples were grouped by site and placed in sturdy plastic containers. Some preliminary sorting of heavy fractions was carried out in the field, but more than half of the heavy fractions were transported to North America for analysis there.

While the bucket method produced excellent results, it has several drawbacks. This method was viable because running water, only recently installed in the modern-day village of Tumbe, was piped in to the dig house/laboratory. For locations on Pemba without running water I suggest floating in a stream or in the ocean if samples can be rinsed adequately prior to drying. For example, during flotation carried out in 2005 by Hajj Mohammed Hajj in southern Zanzibar Island, fresh water was not available so ocean water was used, with great success (Hajj M. Hajj, personal communication, 2005).

The bucket method is also physically taxing, and I received much help with the physical labor of flotation from local Pembans (Selim Hamadi and Hamisi Jeshi) and numerous students and staff on the project. Hajj Mohammed Hajj (HMH) constructed a flotation machine out of local materials, but this commendable device was not operational during the field season. University of Dar-es-Salaam field school students were all introduced to bucket flotation during their participation in laboratory exercises in 2004, and it is hoped that this generation of archaeologists will include flotation in their archaeological methods. Having witnessed the success of flotation first hand through the Pemba Archaeological Project, the head of Antiquities for Zanzibar, Dr. Abdurahman Juma, has subsequently added archaeobotanical recovery to his archaeological operations where possible, beginning with the aforementioned project at Jambiana in southern Zanzibar.

Archaeobotanical Methods - Laboratory

Over 550 flotation samples were recovered during the two seasons of fieldwork on Pemba Island, and of these, approximately 90 samples were chosen for analysis on the basis of context (domestic, feature, midden), time period, site, and level of disturbance. Only light fractions are included in the analysis presented in this thesis. One heavy fraction from each site was sorted fully above 1.4 mm as an independent assessment of the efficacy of the recovery method, but the botanical component of heavy fractions were minimal, and included only wood and some of the denser fruit/nut fragments.

Figure 4.1. First stages of bucket flotation sequence: (a) adding water to buckets (note three buckets for one 10 L sample, operated by Matt Pawlowicz, then a graduate student at the University of Virginia; (b) Hamisi Jeshi (Tumbe resident) agitating sample while filling bucket with water; (c) Selim Hamadi (Tumbe resident) carefully pouring flot and water through light fraction sieve; (d) light fraction sieve after final pouring, showing abundance of charred plant material in addition to some roots.

Laboratory Analysis

Archaeobotanical samples were analyzed according to the standards set out in the Paleoethnobotanical Laboratory Guide (Fritz 1995). Dried flotation samples were sieved for ease of viewing using the following standard geological screen sizes (in mm): 4.00, 2.80, 2.00, 1.40, 1.00, .71, .50, .35. All materials 2.00 mm or larger were completely sorted, and all material under 2.00 mm was scanned for seeds. Wood, coconut shell and fruit/nut fragments were pulled (isolated for counting/weighing) only above 2.00 mm; seeds were pulled from all levels, except for cotton which was pulled only at or above .71 mm. Presence of materials below these cut-offs was noted for inclusion in ubiquity measures. All plant remains ≥2 mm were tallied by weight and number; estimates of minimum seed number (SNE) were also made in the case of fragmented seeds such as cotton and beans. This was done by counting the number of funicular seed caps (cotton) or dividing the number of cotyledons by two and rounding up (Fabaceae). In several instances seed number estimates were reached by dividing the weight of an assemblage of seeds by the weight of an average seed, as determined by weighing 100 whole seeds and dividing by 100. Data was recorded on information sheets during analysis and entered into a Microsoft Excel spreadsheet for analysis and manipulation.

Figure 4.2. Heavy fraction recovery and other activities in the archaeobotany field lab on Pemba: (a) flotation station with light fraction and heavy fraction sieves, operated by Hamisi Jeshi (left) and Salim Hamadi (right); (b) heavy fraction after wet sieving; (c) flotation machine designed by HMH that had limited utility because it was too top-heavy, shown with Hamisi Jeshi (left), the author (center) and Salim Hamadi (right); (d) flotation samples drying in mesh under cover of hut (*kibanda*) roof; (e) preliminary sorting of heavy fraction in the field by Edward Pollard, then a British Institute in East Africa graduate student.

Samples were examined with a stereoscopic light microscope (Olympus SZ 61TR) up to 63x magnification. Digital images of select specimens were taken using a Nikon Coolpix 4500 attached directly to the third ocular of the microscope. Some specimens were chosen for scanning electron microscopy (SEM), which was done at the Institute of Archaeology at

University College London and in the Biological Sciences Department of Washington University in St. Louis, where I was assisted by technician Michael Veith. Three specimens were submitted for accelerator mass spectrometry (AMS) radiocarbon dating by Beta-Analytic, Inc.

Identifications were made using several seed reference guides (Anderberg 1994; Berggren 1969, 1981; Holm 1977; Martin and Barkley 1961; Montgomery 1977; NIAB 2004) in conjunction with local floral guides (Dharani 2002; Koenders 1992; Williams 1949). Archaeological seeds were also compared to modern and ancient comparative materials housed at several institutions including: Archaeobotany Laboratory, Institute of Archaeology, University College London; Paleoethnobotany Laboratory, Department of Anthropology, Washington University in St. Louis; the Institute of African Archaeology and Archaeobotany, Goethe University, Frankfurt; and the Archaeobotany Lab, Department of Archaeology, Simon Fraser University. Many of the materials used for comparison were collected in the field by the author using the methods outlined below in the section entitled "Botanical Comparative Collection".

Data Analysis

Data were converted to ratios for comparison between sites following practices advocated by Miller (1988) and Popper (1988). Most of the data from non-seed botanical categories (e.g., wood, coconut, unidentified fruit/nut, and parenchyma) were converted to density ratios, which are the weight of the given item per liter of original sediment floated (g/liter). Seeds were also converted to density ratios, reported as the number of seeds per liter of sediment floated (#/liter). The need to use two different ratios is due both to the relative difference in abundance between these two categories (most samples contained more than 100 pieces of wood but less than 20 seeds total) and the negligible weight of most seeds, even in large quantities. Density and weight measurements are reported to two decimal places or, more rarely, to three decimal places where warranted – for example, when comparing meaningful differences between two values less than .01.

Data were manipulated using several functions within the Microsoft Excel spreadsheet program, which was also used to create the graphs and charts found in Chapters 5 and 6. In Chapter 7, the diversity of seed taxa within each site was calculated using the Shannon Index (H), which takes into account both species frequency and evenness of distribution (Lloyd et al. 1968). The Shannon index is calculated as:

$$H = - \sum p_i \ln p_i$$

where p_i is the proportion of taxon i to the total number of taxa. This is multiplied by its natural log, and H is the negative of the sum of $p_i \ln p_i$ for all plant taxa at a given site.

Botanical Comparative Collection

Archaeobotanical analysis of eastern African archaeological sites is limited by the lack of comparative materials available for analysis (Wetterstrom 1991). Thus, comparative botanical material was collected during fieldwork on Pemba Island, including important crop species, common weeds of cultivation, and environmental indicator species. Permission from landowners and the Zanzibar Department of Archives, Museums and Antiquities was obtained prior to field collection. When crops were collected from fields, farmers were duly compensated.

Plants were collected on several occasions in 2002 and 2004 from agricultural contexts, home gardens, or wild stands using procedures outlined in standard reference texts (Cotton 1996; Martin 1995) as advocated by the Missouri Botanical Garden (Leisner 2005). No transects or grid collection procedures were followed, as these more intensive methods were beyond the scope of the study presented here. The geographic bearings of each collecting locale were determined using a Garmin GPS receiver, and the terrain, soil conditions, and surrounding flora were all noted. Digital images of each specimen and its environs were taken using a Nikon Coolpix 4500. Only mature specimens (i.e., those bearing flowers or fruits) were collected, and individuals from communities were chosen over specimens occurring alone. Whole herbaceous plants were taken whenever possible, while representative portions of shrubs, trees, and other larger plant forms were taken in quantities sufficient to make five herbarium sheets. Plants were placed in plastic bags and brought back to the laboratory for immediate pressing whenever possible. On those occasions when there was a delay between collecting and pressing, plants were kept moist during transport using water droplets sprayed into the bag.

Plants were then arranged onto small newspaper sheets and placed in a plant press for drying. The press was dried without any electric heating by placing the press on a table exposed to wind but not rain or sun. Some thicker fruits did not dry sufficiently, particularly large mangrove tree fruits, and were treated at the National Herbarium of Tanzania to prevent mold damage. In the future, more effort will be expended to dry specimens quickly, including placing the press over a fire or in direct sunlight (C. Kayombo, personal communication 2004). Herbarium specimens were deposited at the National Herbarium of Tanzania (NHT), located close to Arusha, and were subjected to deep-freezing there for decontamination with the assistance of Canisius Kayombo. Herbarium specimens were then separated and one copy of all taxa collected was sent to an herbarium in Zanzibar, one copy remained at the NHT, and the remaining copies were shipped to the Missouri Botanical Garden (MO). Once at the MO, specimens were identified by myself and Roy Gereau (REG), and were entered by REG into the Tropicos Database, where they can be searched online at: http://mobot.mobot.org/W3T/search/vast.html.

Duplicates of seeds from herbarium specimens remain with the author, and unwent additional decontamination using a -80 degree Celcius freezer at Dr. Jim Johnson's laboratory at the University of British Columbia. The collections (SCW 1-33) are listed in Appendix B. Nomenclature follows guidelines suggested by the Angiosperm Phylogeny Group (APG 1998; 2003; Stevens 2001 onwards).

Ethnobotanical Data Collection

Six individuals and their families provided information about agricultural practices. Most ethnobotanical data was collected through unstructured interviews and participant observation on an opportunistic basis, but several structured interviews were conducted with local farmers. All participants were informed of the nature of the research, the identity of the researcher and assistants, and the voluntary nature of their participation. Interviewees were paid the equivalent of one-third day's pay for each interaction, with the exception of individuals contemporaneously working for the Pemba Archaeological Project, who were already receiving daily wages. This study received research clearance through the Human Subjects Review Committee at Washington University on the conditions that participation was voluntary and that written or verbal consent was obtained.

The majority of ethnobotanical information collected focused on crop processing activities, cooking, storage, and other facets of economic plant production. On one occasion I participated in rice (*mpunga*) processing activities and took samples of the remains of each step in the production sequence, following the examples of Hillman (1984) and D'Andrea and Mituku (2002). On a separate outing, I visited a farmer and his family on the Micheweni peninsula, an area known for its red, rocky soils and the sorghum farming associated with it. One traditional healer was also interviewed to determine the availability and economics of the plants she relied upon most heavily. Due to the sensitive nature of medicinal plant information, and the negative connotation associated with witchcraft on Pemba, I did not pursue this line of investigation further. In all instances, participants' comments were documented on digital images and in field notes. Sound recordings were refused and thus did not comprise part of the interview process, but translations were made on-site through the assistance of H.M.H. and a University of Dar-es-Salaam student.

When inclement conditions prevented laboratory staff from conducting flotation, I often took the opportunity to ask individuals questions concerning economic plants. These opportunistic ethnographic data collection episodes ranged from informal question/answer periods to lengthy structured interviews when time permitted. Through such impromptu interviews, I learned much about coconut plantations and ownership systems, crop storage practices, and the economics of farming and fishing in Tumbe as experienced by three local male crew members.

Conclusions

Three separate but complementary data collection methods were used to obtain information about ancient and modern plants and their uses. While the recovery of archaeobotanical remains was the prime goal of this research, identification and interpretation of these plant remains is limited by the incomplete botanical comparative materials and crop processing information available for this region. Therefore, ethnobotanical data and modern reference materials were collected as an important part of this study. Time and resource limitations precluded extensive ethnobotanical data collection, but future collecting of plants and ethnobotanical data is key to future success in the study of ancient plant use in eastern Africa.

CHAPTER 5: ARCHAEOBOTANICAL RESULTS FROM A.D. 600-1000 ON PEMBA ISLAND: TUMBE & KIMIMBA

Introduction

The reporting of the archaeobotanical assemblages on Pemba has been divided between two chapters. This was done to: a) facilitate comparison of contemporaneous town and village settlement data and b) to break down the large data set into manageable portions. In this chapter I present the archaeobotanical findings from two first-millennium A.D. settlements: Tumbe and Kimimba. Some of the archaeobotanical data from Tumbe has been previously published in Walshaw 2010, however this is the first time the dataset has been available in its entirety. As described in Chapter 3, the earlier levels at Tumbe show the presence of a large seventh to tenth century town comprising earth-and-thatch dwellings, while Kimimba represents a small village occupied between the eighth and tenth centuries. The upper, Mazrui (eighteenth to nineteenth centuries) component of Tumbe is reported separately in Chapter 6 following the description of plant remains from Chwaka and Kaliwa.

A note about excavation terms: Pemban contexts reported here are designated by a unique bag number reflecting an individual locus in 3-dimensional space. Units were dug in natural layers, which were vertically subdivided into arbitrary 10-cm levels if sufficiently deep. In 2002 I took 10-20 liter archaeobotanical samples from every level containing cultural materials at Tumbe, Chwaka, and Kimimba; in 2004, I scaled back and took 10 liter samples from the first level of every layer at Tumbe and Chwaka. The majority of the archaeobotanical data is reported in densities to maximize comparability between samples of different sizes.

Archaeobotanical findings are discussed in terms of the most precise botanical category possible. Thus, while taxonomic identifications were not made on wood fragments or the majority of fruit and nut pieces, well-preserved remains of known taxa, predominantly African and Asian domesticates, were identified to species, and even subspecies. Identification criteria and pictures are provided for each taxon or botanical category found in the Pemban archaeobotanical record. Common names are used throughout, with Latin names used (in italics) where appropriate (see Appendix A for list of all plant names used in this book in Latin, English, and Swahili).

TUMBE – Period 1 (Tana level; A.D. 600-1000) n=23

Twenty-three archaeobotanical samples from Tumbe believed to be associated with the seventh to tenth century occupation are reported, totaling 262.6 liters of floated sediment. Two of these samples represent subsoil, which was sampled to see if overlying deposits percolated downward into sterile layers, and thus to obtain another measure of mixing within the Tumbe deposits. Tumbe deposits are rich archaeobotanically: samples averaged 6.89 grams of flot[1] per liter floated. This is reduced to 3.72 g/liter, however, if we remove bag 1468, a burnt log feature from Unit 63 with an extraordinarily high density of wood at 68.3 g/liter, from the calculation. The portion of the flot that was identifiable to a taxon or category (identifiable flot density) averaged .92 g/liter when the wood feature (71 g/liter) was removed from the calculation. Tumbe Period 1 non-seed data are presented in tabular form, by excavation unit, in Table 5.1 below, and the full data set is presented in Appendix D.

Preservation and Disturbance

All Tumbe strata show evidence of some post-depositional disturbance. On average, Tumbe samples were generally high in uncarbonized materials (averaging .42 g/liter), particularly coconut palm roots and insect eggs. Uncarbonized seeds were present in 87% of samples, typically outnumbering carbonized seeds by several times in Tumbe contexts. These are represented by weedy taxa (*Rumex*, Cyperaceae, Poaceae, Fabaceae, Malvaceae, Asteraceae, Euphorbiaceae, Polygonaceae, Portulacaceae, Amaranthaceae, Arecaceae) most likely intrusive from modern farming activites. Today, Tumbe lands support cassava cultivation, palm groves, and tolerated associates such as *Keetia* (Rubiaceae), and various members of the Cyperaceae (sedge) family.

Wood

Wood charcoal fragments were 100% ubiquitous in Tumbe samples (see Figure 5.1 for Period 1 Tumbe ubiquity data), including one sample that contained wood fragments only smaller than 2 mm, and thus did not contribute to the wood density measure. Wood is also the most common element at Tumbe, by both count and weight. Wood density (see Figure 5.2 for Tumbe density data) shows a pattern similar to and indeed, related to, flot density mentioned above. Taken together the Tumbe samples average 3.24 g/liter of carbonized wood, but if we exclude bag 1468 from consideration, the average wood density at Tumbe drops to a mere .14g/liter. This is still a significant portion of the Tumbe archaeobotanical deposits, and is best interpreted as the burning of wood for fuel.

No attempt was made to identify wood specimens to any taxonomic level (see Figure 5.3 for digital image of archaeological wood from Tumbe). As discussed in Chapter 3, mangrove trees were important sources of fuel and structural materials, and would have been

[1] Flot is the term for the material that floats during water separation, also known as the light fraction (Pearsall 1989:33, 56).

Table 5.1. Non-seed archaeobotanical data for Tumbe Period 1.

Unit	Number of Samples	Liters floated	Wood weight (g)	Parenchyma weight (g)	Root/ tuber #	Coconut weight (g)	Fruit/Nut weight (g)	UNID weight (g)	Uncarbonized weight (g)
14	2	23	5.57	0	0	.80	3.86	.02	8.06
16	3	56	4.17	0	0	.69	.46	0	32.91
47	1	10	0	0	0	.02	0	0	5.95
50	1	10	.68	0	0	.03	0	.01	1.35
56	1	12	3.65	0	1	0	0	.13	4.76
60	4	35	.80	.02	2	.20	.20	.03	10.8
63B	3	23	13.43	0	1	.98	.76	.41	8.57
63D	1	6	409.80	0	0	.02	.82	0	2.79
64	5	72	5.01	.20	3	.22	.53	.26	23.7
OP4E	2	16	3.60	.02	0	.20	.20	.02	8.5
Site total density (g/liter)			**3.11**	**.01**	**---**	**.14**	**.02**	**.01**	**.42**
Site Total Ubiquity			**100%**	**26%**	**17%**	**87%**	**74%**	**83%**	**100%**

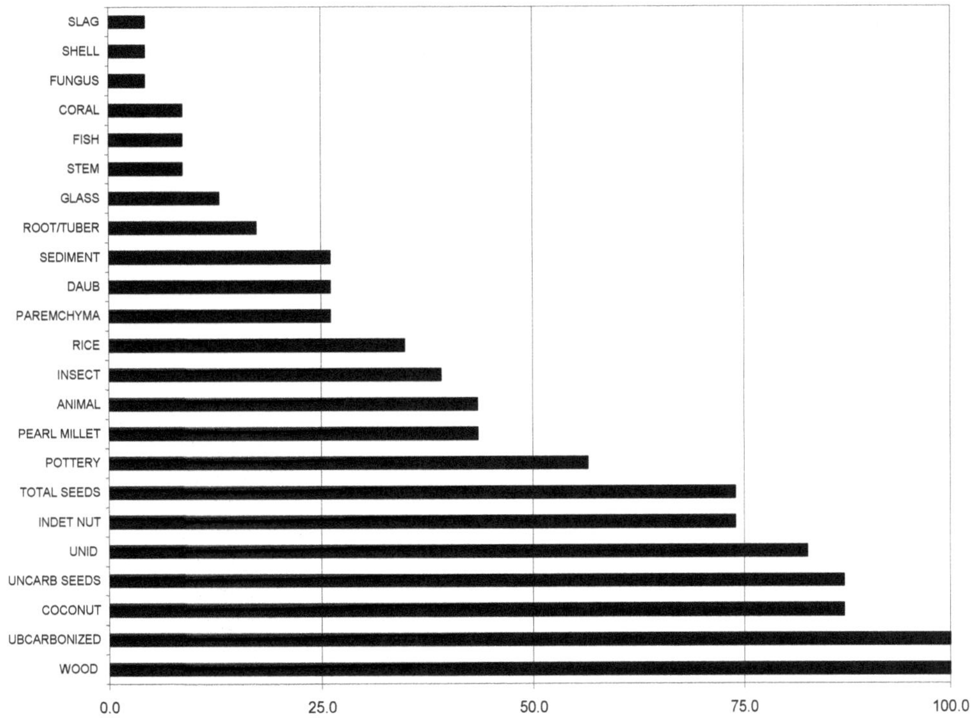

Figure 5.1. Ubiquity (percent frequency) of materials found ≥ 2 mm at Tumbe.

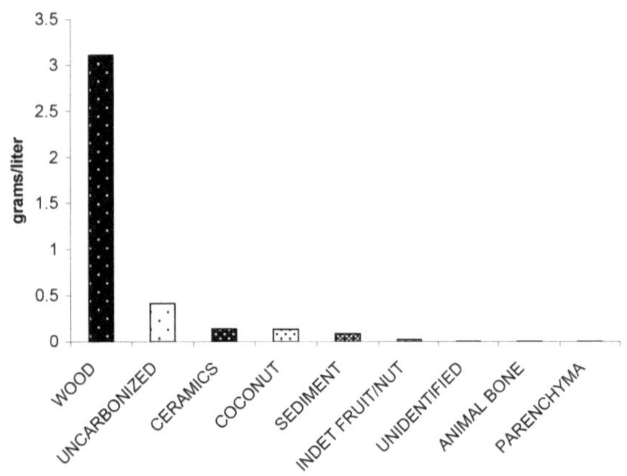

Figure 5.2. Density of materials found ≥ 2 mm at Tumbe.

Figure 5.3. Archaeological unidentified wood fragments from Tumbe. Scale is in mm.

available to Tumbe residents. There are, however, many other wood species available on Pemba, as well as alternative fuel sources, so the identification of wood sources would contribute significantly to our understanding of resource exploitation on Pemba. The frequent presence and excellent preservation of wood at Tumbe makes it a good candidate for further study.

Storage Tissues

Parenchyma, defined as relatively undifferentiated plant tissue such as that found in underground storage organs (Hather 2000:1), was identified as carbonized plant material with undifferentiated cellular structure and no discernable secondary vascularization or covering. Possible sources of the archaeological parenchyma at Tumbe include, among others, yam tubers, banana fruit flesh, and taro tubers. No effort has been made towards taxonomic identification of Tumbe's parenchymal fragments, but SEM imaging of a sample of archaeological tissues was conducted as a first step toward making identifications.

Parenchyma was found in only eight samples from Tumbe (26.1% ubiquity). Interestingly, this seems to be concentrated in Unit 64 (80% ubiquity), a deposit with large amounts of pottery sherds and bead-grinders. In particular, all samples from Layer 5 within Unit 64 contained parenchymal tissue, including three feature samples and one composite sample. Parenchyma is a light material and is sparsely represented at Tumbe; both these factors contribute to its mean density below .01 g/liter.

Small tubers with skin still attached, some similar to *Cyperus rotundus*, were categorized not as parenchyma, but under the umbrella term "Root/Tuber". Roots/tubers were found in four Tumbe samples (17.4% ubiquity), numbering seven fragments in total (see Figure 5.4 for image). The mean density was far below .01 g/liter, therefore this category is not represented in Figure 5.1.

Figure 5.4. SEM image of archaeological fragment of underground organ from Tumbe. Scale is in micrometers (µm).

Coconut (Cocos nucifera) *and Relatives*

Coconut endocarp ("shell") fragments were found in most samples from Tumbe: 18 samples (78%) contained identifiable coconut; 2 (8.7%) more contained possible coconut. While coconut ubiquity is high at Tumbe, its frequency is comparatively low: average density of coconut was .01 g/liter, which is less than 1/10th the density reported for wood. Coconut fragments were typically quite small, less than 4 mm in size, so identification was based on patterns exhibited by experimentally charred and fragmented coconut specimens. The small fragments of coconut that frequent the archaeobotanical record tend to show minimal to no curvature and display a

Figure 5.5. Archaeological coconut fragments from Tumbe. Scale bar is in mm.

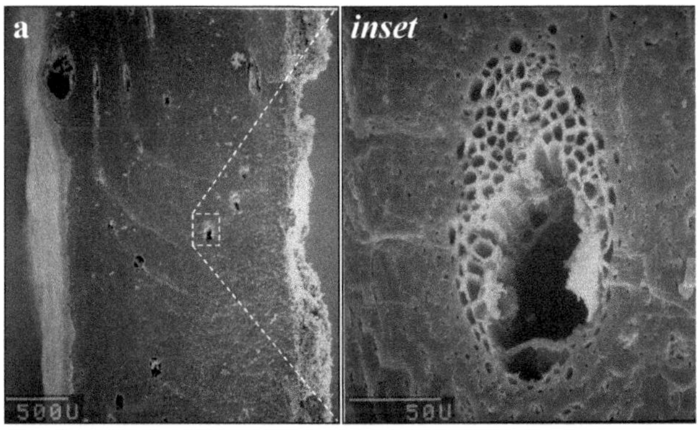

Figure 5.6. SEM images of modern coconut endocarp: (a) endocarp in cross-section; (inset) close-up of vascular pore showing xylem (water conducting) tissues. Scale bars are in micrometers.

blocky profile dotted occasionally with small vascular pores (see Figures 5.5, 5.6). Both the exterior and interior surfaces have a smooth aspect, infrequently with linear ridges and indentations or, more rarely, showing evidence of the "eye" (germination pore) of the coconut. Finally, while Ohler (1999a:21) reports fresh coconut endocarp thickness between 3 mm and 6 mm, measurements taken on mature and immature East African Tall coconut shells available on Pemba Island range between 2.1 mm and 5.8 mm, the thickest portion being on the vein (see Table 5.2 for details). One unripe coconut (*dafu*) shell was measured to between 2.4 mm and 4.8 mm in thickness. Experimentally charred East African Tall coconut (n=1) ranged between 1.8 and 2.3 mm, indicating a rate of shrinkage in mean endocarp thickness approaching 50% upon carbonization. Experimental charring of coconut originating from the Dominican Republic obtained at Shnuck's supermarket in St Louis resulted in endocarp thicknesses between 2.0 and 4.0 mm, on average 28% smaller than the fresh endocarp. Further experimentation with the shrinkage rates of mature and immature coconut endocarp are needed to solidify these preliminary findings, but the data support using 2.0 mm minimum thickness for identifying coconut endocarp archaeologically.

Table 5.2. Fresh coconut endocarp thickness (in mm) of East African Tall variety from Pemba Island (Mature 1 – Immature 1), experimentally charred East African tall endocarp (Charred 1) and one experimentally charred store-bought specimen originally from the Dominican Republic (Charred 2).

Sample	Minimum	Maximum	Mean*
Mature 1	3.0	5.8	4.2
Mature 2	2.9	5.8	4.3
Mature 3	3.7	5.6	4.6
Mature 4	3.0	5.2	3.7
Mature 5	2.1	5.4	3.3
Immature 1	2.4	4.8	3.2
Charred 1	1.8	2.3	2.1**
Charred 2	2.0	4.0	2.8

* of 8 measurements
** of 12 measurements on charred fragments

A fragment of coconut endocarp from Tumbe was dated directly using AMS radiocarbon technology (Beta-202852) to 1180 +/- 40 BP (cal. AD 770-970 to two sigma range; see Appendix D for table of radiocarbon dates). This appears to be the earliest directly dated coconut specimen from the coast of eastern Africa, and confirms the presence of coconut from the earliest archaeological evidence of human habitation on Pemba Island. This finding is not the earliest coconut in Africa, however; coconut shell was found in Roman deposits from Berenike in Egypt, dated by association to the first and second centuries A.D. (Wendrich et al. 2003). Cappers considers the Berenike coconut to be of Indian origin and destined for Roman consumption (Wendrich et al. 2003). The Tumbe coconut date also seems contemporaneous with Hoffman's (1984) report of a few coconut fragments from ninth and tenth century deposits on the Comoros Islands, and likely signals the advent of coconut cultivation in eastern Africa.

Coconut Evolution, Dispersal, and Cultivation History. *Cocos* is a monospecific genus within the Cocoeae. Recent molecular work (Gunn 2004) shows that the coconut is an ancient lineage (circa 23 million years) most closely related to Neotropical palms, and more distantly related to the African oil palm, *Elaeis guineensis*. These new data counter the common view of a Southeast Asian origin for the coconut by suggesting that *Cocos* originated in South America and dispersed eastward and southward through natural, and more recently human, dispersal mechanisms to reach its current pantropic distribution. Two subspecies have been described: *C. nucifera typica,* containing the tall palms, and *C. nucifera nana,* to which the dwarf palms belong. Dwarf palms are typically much shorter and bear smaller, more triangular fruits, than the *typica* varieties, which are supported by taller, straighter stems and produce larger, more rounded fruits. On Pemba today the East African Tall variety dominates, with some indigenous Borassus palm (*Borassus aethiopicum*) cultivated. Evidence for oil palm or other members of the coconut tribe, Cocoeae, were found in other Pemban archaeological contexts, and these will be discussed further in Chapter 6.

There is some debate as to how the coconut arrived in eastern Africa: by human dispersal through contact with Asian trade, or naturally across the ocean. This is a common concern with this important palm, which today enjoys a wide tropical distribution mostly owing to human dispersal (e.g., Ward and Brookfield 1992). Our ignorance of the biogeographic history of coconut in eastern Africa is mostly due to inadequate sampling in this region, as opposed to the New World palms which have been the subject of intense study, both of modern and archaeological specimens (Morcote-Rios and Bernal 2001). Some researchers credit human dispersal with introducing coconut to eastern Africa (Muir 1937), some argue that ocean dispersal alone could have brought coconuts there (Sauer 1967: 45-47), and still other remain equivocal (Dennis and Gunn 1971). Coconut fruits are unusually buoyant and can float for up to 110 days, with a projected travel distance of 3000 miles if floating on the faster of the globe's ocean currents (Dennis and Gunn 1971). Moreover, coconuts are dioecious, and thus one coconut fruit is sufficient to start a new viable colony. Dennis and Gunn (1971:411) assert that "the coconut would appear to have all the qualifications of a plant especially adapted to coral islands"; if this is indeed the case, Pemba would certainly have provided a hospitable environment for drifting coconut fruits. But the trading of coconut between India and Africa by the Roman period suggests that the coconut was known to Red Sea ports, and likely was introduced south along the coast from these northern ports during the first millennium A.D.

The cultivation history of coconut is complicated by several factors. First, wild and cultivated types cross-fertilize, and the resulting complex histories of introgressive hybridization between natural and introduced phenotypes have made it difficult to identify a geographic origin on the basis of molecular analyses. Furthermore, coconut fruit morphology is adapted to long-distance dispersal through drifting on ocean currents, and, as discussed above, there has been much discussion over whether coconut's pan-tropical distribution is more the result of human introduction or natural dispersal (Dennis and Gunn 1971; Harries 1992; Sauer 1967). Harries (1978; 1981) has developed the most widely accepted explanatory framework of coconut morphology and taxonomy to date. He describes two adaptational extremes for the tall coconut: the niu kava type is adapted to oceanic dispersal by means of an angular, thick-husked fruit, whereas the niu vai type has developed a rounded, thin-husked fruit in response to human selection. Most coconut varieties existing today are intermediate between these two extremes, and Harries uses such characteristics as germination time (shorter in niu vai types) and husk size relative to fruit size (smaller in niu vai types) to categorize extant varieties. Endocarp thickness may also have decreased in response to human selection, and thus be observable in the archaeological record, but comparative metric studies involving the coconut endocarp have not been published to date. This is an intriguing line of inquiry for future investigation.

Many portions of the coconut palm were undoubtedly used in the past, but typically only remains of charred fruit shell (endocarp; see Figure 5.7) are present at archaeological sites (but note presence of coconut shell and husk at Berenike). Potentially, the fibers of the leaf used as construction materials may survive in the archaeological record, but for the purposes of the present discussion, only endocarp morphology is considered here. Coconut shell is reported to have been an important fuel source in the past, and the abundance of coconut shell in the archaeobotanical record is likely owed to preservation of endocarp by intentional burning or discarding as refuse after extraction of the fluids and meat.

The coconut is one of the most esteemed of the tropical trees, owing to its many important food and fiber products (Ohler 1999b) . The oil-rich white endosperm commonly referred to as "meat" can be eaten fresh, pressed for oil, or dried into a storable form, called copra. Immature fruits yield a watery endosperm drink called *dafu* in Swahili. The endocarp, colloquially referred to as the coconut "shell", can be used as a container, fashioned into tools and crafts, or burned as fuel. The fibrous layers of the mesocarp, or "husk", can be made into a rope called coir, and the leaves and leaf-bases serve as important thatching and construction materials.

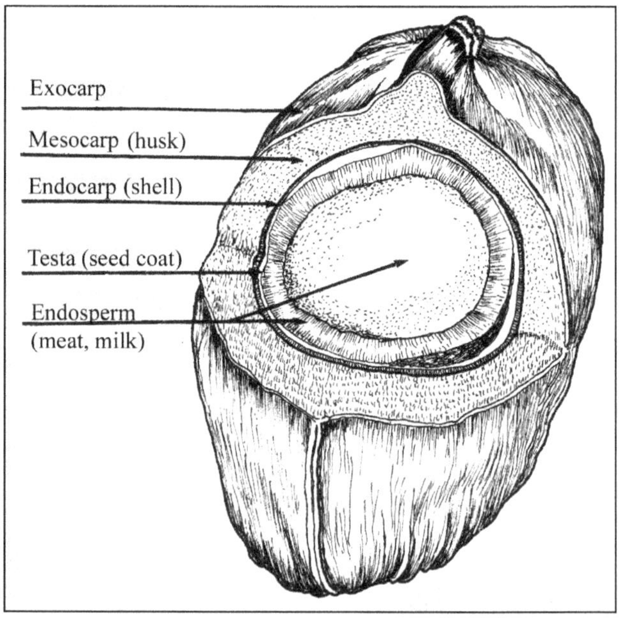

Figure 5.7. Coconut fruit morphology. Adapted from Gunn and Dennis (1976;183 Figure 78).

Coconut was one of the most importante conomic plants in the ancient Indian Ocean world; indeed, the seventeenth century French traveler Pyrard noted: “Maldive ships were built, fitted, equipped, and provisioned entirely from these trees and then loaded with cordage, mats, sails, oil, wine, sugar, and other products, all derived from the same palm, for trade with the East and Arabia” (Gray, ed. and transl. 1887-90 Pyrard, in Sauer 1967:42).

Other Fruit/Nut

At Tumbe, other fruit and nut casing fragments that could not be assigned to coconut were collapsed into one category, and future research is planned to determine the identities of these fragments. Items categorized as fruit/nut casings are typically dense tissues with some internal cavities or external coverings, or sometimes parenchymal tissue that has come apart from the remainder of the fruit casing (see Figure 5.8). In one case, a whole small fruit was recovered. Most (74%) of Tumbe Period 1 samples contained some unidentified fruit and nut fragments, with an average density of 0.02 g/liter.

Possible candidates for the fruit/nut category include castor fruit casing (*Ricinus communis*), local African trees such as the baobab, fig (*Ficus* sp.), and other palms (e.g., oil palm, Borassus palm), as well as introduced species, such as breadfruit, jackfruit, mango, and betel nut, among others.

Figure 5.8. Unidentified fruit/nut fragments from Tumbe. Scale is in mm.

Other Botanical Items

Most samples (n=18; 83%) contained botanical fragments that could not be identified to a taxonomic category, and hence were labeled “UNID BOT” (unidentified botanical). On average, Tumbe samples contained unidentified botanical material at a density of less than .01 g/liter of sediment floated.

Seeds & Chaff - Domesticates

Tumbe samples analyzed to date have produced a total of 375 seeds, excluding chaff fragments from rice and pearl millet. Seventeen samples (74%) contained seeds, with an average of 22.1 seeds per sample, or 3.7 seeds per liter of sediment sampled. Two samples from Unit 14 were responsible for 70% of all seeds from Tumbe owing to the abundance of pearl millet grains found there. Most of the seeds from Tumbe are domesticates: pearl millet, rice, beans, and cotton occur most frequently (see Figure 5.9, Table 5.3) A few common weeds of agriculture are found as well, although in low frequency and abundance, including dock, amaranth, and members of the spurge and poppy families.

Pearl millet (Pennisetum glaucum L. Br.). Tumbe contains the vast majority of pearl millet grains and chaff found throughout the Pemban archaeobotanical materials (see Figure 5.10 for pearl millet images). This is interesting considering the infrequent and scarce findings of pearl millet at the other Pemban sites discussed below. Tumbe deposits contained a total of 174 grains of identifiable pearl millet as well as 104 grass grains likely to be pearl millet but distorted such that key characters were missing or unrecognizable. Domestic pearl millet grains are club-shaped with a round cross-section and distal dimple (D'Andrea et al. 2001). While pearl millet is best identified on the basis of grain shape, it can also be distinguished metrically from sorghum by shorter overall length (<3 mm fresh), and a thicker, longer embryo that is 2/3-3/4 the length of the grain. Small amounts of the tiny pearl millet involucre bases (13 total) were also found in a few samples, but only from those contexts where pearl millet grains were well represented. Ubiquity of pearl millet at Tumbe is 43%, with 17.4% of samples containing pearl millet chaff as well.

Table 5.3. Seed and chaff representation by unit at Period 1 Tumbe.

Unit	14	16	47	50	56	60	63B	63D	64	OP4E	Total
Sample size (n=)	2	3	1	1	1	4	3	1	5	2	**23**
Liters floated	23	56	10	10	12	34.6	23	6	72	16	**262.6**
Pearl millet grains	144	0	0	1	18	0	4	0	5	2	**174**
Poaceae like pearl millet grains	96	1	0	0	5	0	0	0	0	1	**103**
Pearl millet chaff (involucre bases)	10	0	0	0	2	0	0	0	1	0	**13**
Finger millet grains	1	0	0	0	0	0	0	0	0	0	**1**
Sorghum grains	0	1	0	0	1	0	0	0	0	1	**3**
UNID millet	4	0	0	0	0	0	0	0	0	0	**4**
Poaceae like Triticeae (wheat/barley)	1	2	0	0	0	0	0	0	0	0	**3**
Rice grains	2	0	0	1	0	1	0	0	0	0	**4**
Rice chaff (rachillae)	4	14	0	0	0	0	0	0	1	1	**20**
UNID Poaceae	0	0	0	0	0	0	0	0	1	0	**1**
***Vigna* sp.**	0	0	0	0	0	0	4	0	0	2	**6**
UNID Fabaceae	0	0	0	1	0	0	0	0	0	2	**3**
Cotton	1	1	0	0	1	0	0	0	6	2	**11**
***Rumex* sp.**	8	0	0	0	0	0	0	0	1	1	**10**
Polygonaceae	0	1	0	0	0	0	0	0	0	0	**1**
Euphorbiaceae	0	0	0	0	0	0	1	0	0	0	**1**
Portulacaceae	0	1	0	0	0	0	0	0	0	0	**1**
UNID A (Papaveraceae)	0	1	0	0	0	0	0	0	1	0	**2**
Sesame	0	0	0	1	0	0	0	0	0	0	**1**
UNID seeds	2	0	0	0	3	6	12	5	12	0	**40**
Unit Seed and Chaff Total	**273**	**22**	**0**	**4**	**30**	**7**	**21**	**5**	**28**	**12**	**402**

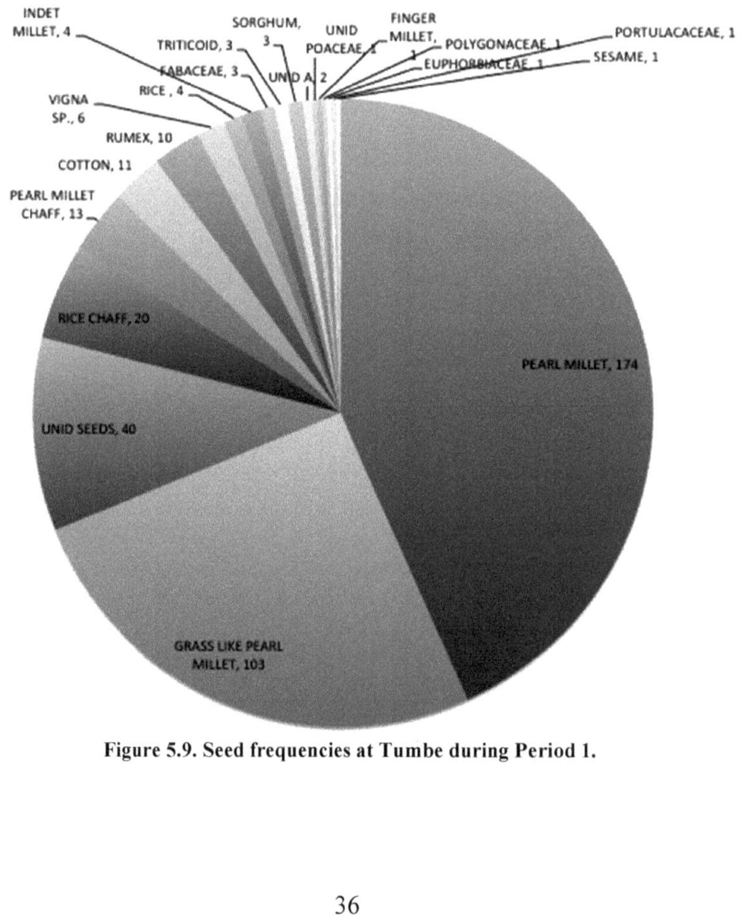

Figure 5.9. Seed frequencies at Tumbe during Period 1.

One pearl millet sample from Tumbe was submitted for AMS radiocarbon dating (Beta-202827), which resulted in the earliest known date for the site of Tumbe: 1370±40 B.P., or a calibrated date of A.D. 620-700 (95% probability; Walshaw 2010). This grounds the earliest occupation of Tumbe well within the 7th century AD, and provides a minimum date for the spread of pearl millet agriculture to the eastern coast of Africa. Interestingly, pearl millet was not among the cultigens reported at Axum, an earlier (mid-first millennium B.C to mid-first millennium A.D.) urban site in Ethiopia (Boardman 1999), and it appears at sites on the Zimbabwe plateau after 1800 A.D. (Jonsson 1998).

Commonly known as bulrush millet, pearl millet is a common crop in West Africa and India, with sporadic cultivation across eastern Africa's drier regions under 1200m elevation. It has a distinctive condensed floral spike, bearing seeds that are pale grey to beige, and shiny (Acland 1971:27-28; NRC 1996:89) . The grains may be ground into rough flour for making porridge (*uji* in Swahili) or beer (*pombe*), preparations which may lead to a decreased chance of preservation of identifiable grains in the archaeological record. Pearl millet is a prominent component of archaeobotanical assemblages from West and Central Africa (Amblard and Pernès 1989; D'Andrea and Casey 2002), where it appears in domesticated form at the site of Birimi in Ghana by 1900 B.C. (D'Andrea et al. 2001).

Generally the pearl millet grains at Tumbe are small compared with published measurements of domesticated modern, wild, and shibra (modern x wild) pearl millet (Brunken 1977), but accord with D'Andrea et al.'s (2001) archaeological pearl millet from Birimi (3500-2500 B.P.) (see figure 5.11). Moreover, pearl millet metric data from Tumbe and the later sites of Chwaka and Kaliwa are remarkably similar, with the exception of one large outlier from Tumbe. Since both the Birimi and Pemban specimens display domestic-type morphology (club-shaped grain, circular in cross section [Brunken 1977; Brunken, de Wet and Harlan 1977]) the small size alone cannot be interpreted as indicating a wild-type, but it does require an explanation. Charring is a commonly invoked explanation in considerations of metric aberrance, but that may not apply in this case. Pearl millet grains appear to expand approximately 7-8% in both breadth and width when charred (Abdel-Magid 1989:303), which is consistent with results from other grain charring experiments (Nesbitt 1997:183-186) in which the breadth to width ratio remains unchanged upon charring. The small archaeological specimens may represent immature or under-ripe grains that may have been discarded during processing. When a whole spike of pearl millet is harvested, not all of the grains may be equally ripe, particularly the spikelets at the apex (top), analogous to "tail grain" in wheat or the diminutive kernels on the distal end of a corn cob. At least one of the grains from Tumbe showed a medial curvature to the embryo that may be consistent with immaturity, but most of the grains were lacking an embryo, which is easily lost or distorted when charred. This hypothesis does not stand up to metric testing, however; measurements on basal, middle, and apical grains on a pearl millet spike reveal no clear size trend (see Figure 5.12).

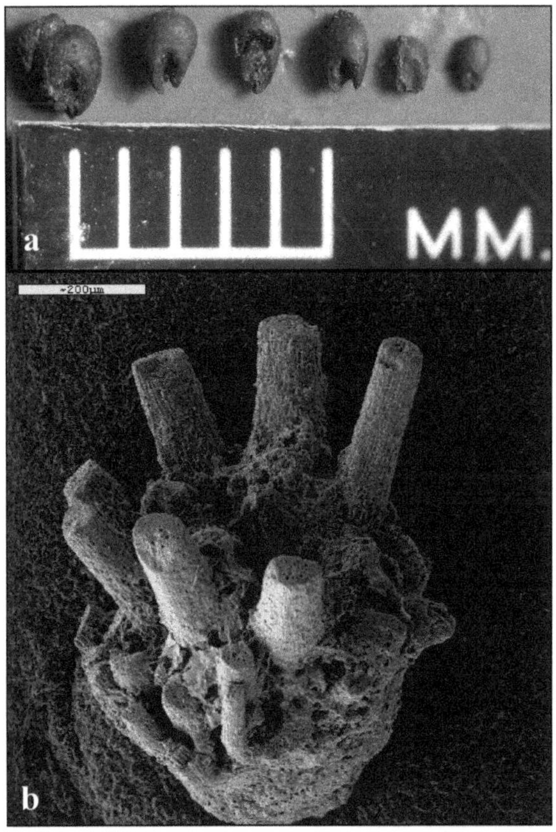

5.10. Archaeological pearl millet from Tumbe: (a) micrograph of grains with scale in mm; (b) SEM of spikelet base with scale in micrometers.

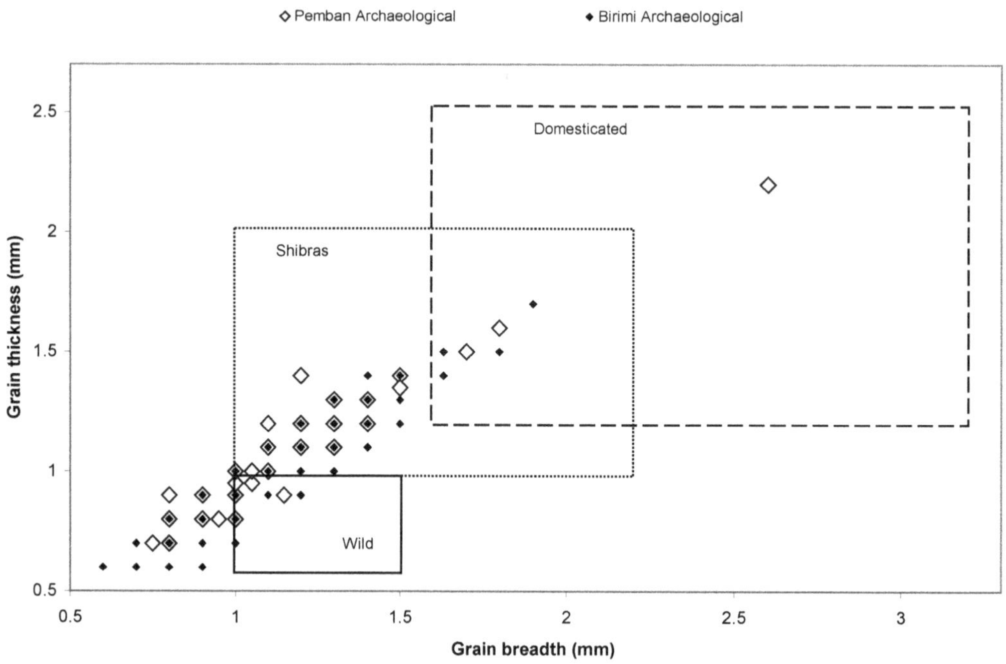

Figure 5.11. Archaeological pearl millet measurements for Pemban specimens (open diamond) and Birimi specimens (from Ghana; data from D'Andrea et al. 2001: 345 Figure 5; filled diamond). Superimposed rectangles indicate range of wild, shibra (intermediate form) and domesticated modern pearl millet (data from Brunken 1977: 165-170).

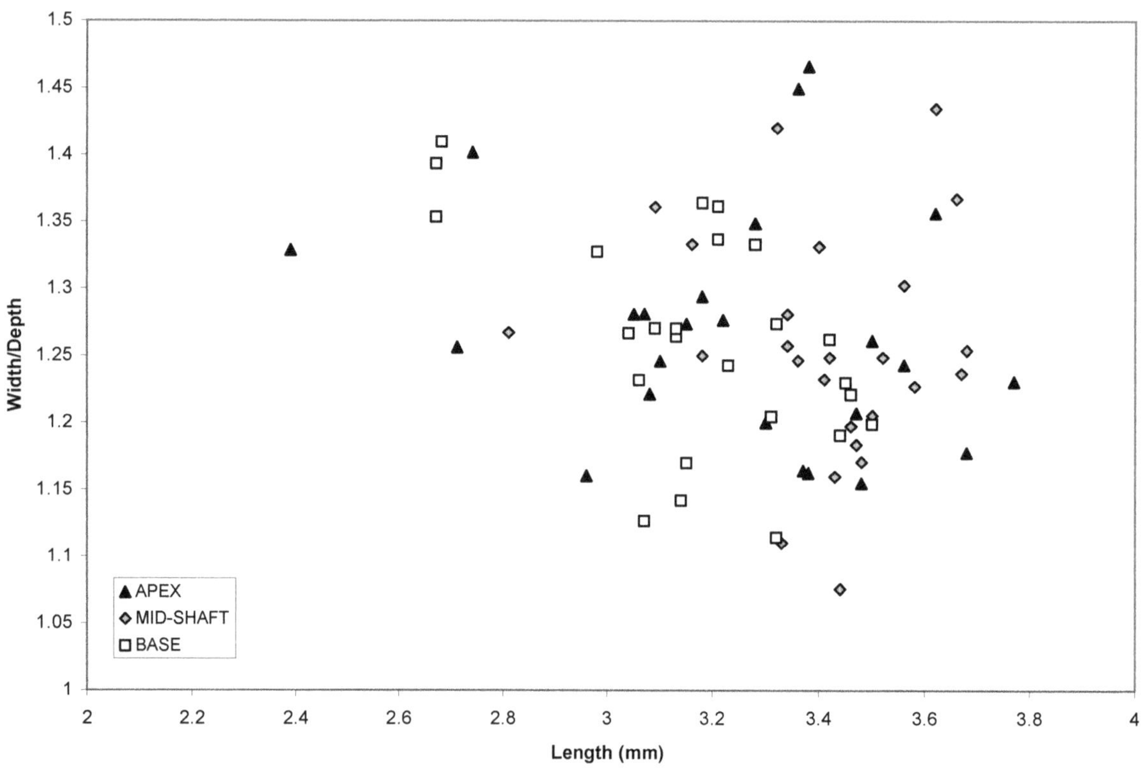

Figure 5.12. Measurements comparing size of grains distributed along the shaft of a modern dried pearl millet spike obtained from Pemba Island.

Basal grains were, overall, on the smaller side, but clearly overlapped in range with grains from the mid-shaft and apex. A second theory concerning the small size of pearl millet grains is based on environment: Pemba's lush landscape is far wetter than the arid woodland savannahs and grasslands that support pearl millet cultivation today and in the past. If pearl millet on Pemba was compromised due to fungal infections, grains may not have attained maximum size, or infected specimens may have been discarded and thus have an opportunity to enter the archaeological record. Yet another explanation may be inter-crossing with the wild progenitor and the formation of an intermediate form, known as a shibra. It remains possible that early Bantu farmers introduced both wild and domestic pearl millet forms when expanding into eastern Africa. Clearly this issue requires more data to be resolved, including more measurements of archaeological specimens, and a greater understanding of the size variation among wild, intermediate, and cultivated pearl millets (D'Andrea et al. 2001).

Finger Millet – Eleusine coracana *ssp.* coracana *L. Gaertn.* One finger millet grain was recovered from Tumbe deposits, resulting in a ubiquity measure of 4.3% (depicted in Chapter 6). Finger millet is a small rounded grain with a very steep embryo and diagnostic pustules in waved lines on the seed coat. Additionally, because of its roundedness (globularity), the finger millet has a round hilum immediately below the impressed rounded embryo, contrasting with (and distinctive from) the hilum's presence on the flattened ventral surface of most other millets (Fuller 1999:C11).

The presence of finger millet on the coast is somewhat unexpected, because this grain is mostly grown in highland African settings today. But finger millet is most likely an eastern African domesticate, even though evidence for the domestication process is absent in the archaeological record. While the earliest archaeological specimens date to 1800 B.C. in India (Portères 1976), recent re-examination of the identification criteria for these reports has revealed inconsistencies and likely misidentifications among putative prehistoric finger millet finds in South India (Fuller 1999, 2003). Cultivated finger millet is interfertile with *E. coracana* ssp. *africana*, a millet indigenous to Africa (Harlan et al. 1976:417), which strengthens the argument for an African domestication. The paucity of archaeobotanical information from East African sites (Wetterstrom 1991) challenges our knowledge of the origins and spread of finger millet cultivation in Africa beyond linguistic interpretations of agricultural histories. Domesticated forms have been recovered at Axum, in Ethiopia, dated to A.D. 500 (Boardman 1999), at Deloraine Farm, Kenya by the eighth century (Ambrose 1984), and in southern Africa by the seventh century (Maggs and Ward 1984; Pwiti 1996).

Sorghum (Sorghum bicolor *ssp.* bicolor). Two poorly preserved sorghum grains were found at Tumbe (8.7% ubiquity). Sorghum is a rounded to slightly club-shaped grain with a shorter, rounder embryo than pearl millet, and is typically larger than the other African millets (see Figure 5.13 for an image of sorghum from Chwaka). According to ethnohistoric reports, sorghum was a Swahili staple, so its relative absence is surprising until one considers that it, like pearl millet and finger millet, was likely ground for use in a liquid (e.g., beer, porridge). As such, it would become charred infrequently, and, if charred, would probably be unrecognizable. Preparation methods and other cultural practices have been reported as contributing to eastern Africa's scant archaeobotanical record (Young and Thompson 1999).

Cultivated sorghums have the largest grains of all the millets[2], with lengths typically exceeding 4 mm (Fuller 1999: C21). Wild sorghum is relatively thinner and narrower, particularly in comparison to the extreme plumpness (verging on roundness) in the highly evolved durra varieties. Cultivated types have wide and ovate caryopses (where the width almost equals the length), embryos that are shallow, wide, and long (1/2-2/3 of the overall seed length) and a projecting, pointy hilum. Charring of most millets, large and small, appears to reduce the size of the grain but increase the depth of the grain at the apex (side away from embryo) through a separation of the dorsal and ventral surfaces which allows the starchy interior material to push through, similar to that observed when maize 'pops' (Fuller 1999:210).

Figure 5.13. Archaeological sorghum from Chwaka. Scale is in mm.

[2] "Millet" is a colloquial name applied to a group of small-seeded grasses that serve as food and fodder. Most millets belong to the sub-tribe Setariinae within the panicoid grass tribe (Fuller 1999:209). Millets are distinctive from other economically important cereals (wheat, barley, rice) by their widely ovate caryopses, lack of ventral furrow, long and wide embryos and rounded hilum.

The sorghum found in the Pemban archaeological record was too distorted to be accurately identified to race, but the concentration of sorghum collected from the archaeological site of Kilwa (stonetown located on the southern Tanzanian coast) may be identifiable to race, should the collection still be available for analysis.

The most reliably identified, but not yet AMS radiocarbon dated evidence, from Indian sites shows cultivated sorghum of the bicolor race appearing in the second millennium B.C. (Fuller 1999:C24). However, wild forms are today restricted to Africa south of the Sahara (Zohary and Hopf 2000:88). At Nabta Playa, Wasylikowa and Kubiak-Martens (1995) found evidence that wild sorghum collection predated sorghum cultivation in southern Egypt at approximately 6000 B.C. Cultivated sorghum does not appear at African archaeological sites until surprisingly late: by 800-400 cal. B.C. at Kawa (Fuller 2004) and after AD 100 at Qasr Ibrim, both in the Sudan (Rowley-Conwy et al. 1997). Sampling may be at issue, however: one should consider the "lack of systematic archaeobotany in the Sudanic savannah/sahel belt between 4000 B.C. and 0 A.D., and that Qasr Ibrim lies too far outside the region in which domestication would have occurred to be useful [in determining the antiquity of sorghum cultivation]" (Fuller 1999:C25).

Based upon the distribution patterns found among extant African sorghum varieties, Harlan and Stemler (1976) have postulated an African domestication and expansion scenario for *S. bicolor*. According to their model, sorghum was domesticated in Africa, roughly in the eastern Sahel zone. *Guinea* sorghum was then developed out of this cultivar quite early in West Africa, and the early *bicolor* variety also moved south to East Africa, where it evolved into the variety referred to as *kafir*. The *durra* race is hypothesized to be Indian in origin, involving an early spread of *bicolor* to India, subsequent mutation, and a transmission back to African, likely during Islamic times (Harlan 1993). Sorghum finds along the Swahili coast, therefore, may be of either the *kafir* or *durra* type, depending upon the route of transmission, and future identifications will hopefully be instructive in this regard.

Rice (*Oryza sativa*). Rice grains are less common than those of pearl millet at Tumbe, although the chaff elements are more frequently encountered than the grains themselves, which is the reverse situation to that of pearl millet. Five rice grains or fragments in total were recovered (seed number estimate [SNE] =5; 17.4% ubiquity). Tumbe also produced 28 rice rachillae (also sometimes called spikelet bases or involucre bases - the part of the plant that supports the single floret), resulting in a ubiquity of 21.7% for rachillae and 30% for all rice elements. Direct dating of the rice at Tumbe awaits the recovery of a whole grain; the rachillae are individually too small to date and a conglomerate date may produce a large error.

The rice at Tumbe is not in sufficiently whole condition to permit confident assignment to either Asian or African rice. African rice, *Oryza glaberrima*, is much more geographically restricted than Asian rice and was likely not cultivated in eastern Africa, although it remains an important crop in West Africa today (Harlan 1993). Moreover, as will be presented in the Chapter 6, rice found at Chwaka is of the short-grained variety likely of temperate origin but grown throughout Asia today. The Tumbe rice may be of a similar variety to the rice described at Chwaka, although the finding of an entire rice grain is required to confirm or reject this hypothesis. Greater consideration of the rice on Pemba is presented in Chapter 6 alongside a discussion of the abundant rice present at Chwaka.

Unidentified Grains and Chaff. Four unidentified millet grains were found in one sample (4.3% ubiquity) at Tumbe. This was from a context rich in pearl millet, but these specimens are most likely not pearl millet, which was confined to the categories "pearl millet" and "Poaceae like pearl millet".

Unidentified grass straw fragments were observed in two of the 23 samples (8.7% ubiquity). Some stems appear to be from rice plants, although accurate taxonomic identifications from rice stems are best achieved through phytolith analysis (Harvey and Fuller 2005). Therefore, conclusive data regarding the presence of rice culm elements at Tumbe are not available at this time. Grass straw was likely a component of earthen architecture, used both as thatch for roofing and temper for mud walls; therefore the presence of unidentified straw at Tumbe could be of architectural origin, rather than due to crop processing.

Vigna *sp.* Three Tumbe samples (13%) produced six beans or bean fragments resembling to the species *V. radiata*, the mung bean, and/or a congener (see Figure 5.14 for an image of a *Vigna* specimen from Chwaka). Fabaceae seeds in general are infrequent among Tumbe samples (17.4% ubiquity), but one sample (bag 1275 in Unit 63B) was responsible for four of the six probable mung beans.

Beans and peas are common features of the Swahili garden and diet today, so their presence is entirely expected. However, the recovery of mung bean, an Indian legume, in the archaeobotanical record at Tumbe is somewhat surprising. One might expect to find the African legumes to be present, including hyacinth bean (*Lablab purpureus*), cowpea (*Vigna unguiculata* ssp. *unguiculata)*, or Ethiopian pea (*Pisum abyssinicum*). Legumes are typically poorly preserved at archaeological sites, possibly because of their minimal processing and wet preparation, so the low frequency of legumes in general at Tumbe follows these expectations.

Cotton (Gossypium *sp.*). Cotton seeds were found in seven Tumbe samples, resulting in a 30% ubiquity for this taxon. Cotton seeds are easily recognizable by their relatively large size, globose oval shape, palisade internal structure to the seed walls, robust small round "cap" or funicular attachment, and the occasionally preserved seed stem (see Figure 5.15). While the Tumbe specimens are too fragmentary to permit a satisfactory subspecies identification, whole specimens from Chwaka resemble most closely the Old World domesticated cotton *Gossypium herbaceum*. I discuss cotton in greater depth when considering the cotton recovered from Chwaka in Chapter 6. The presence of cotton seeds is very interesting, particularly considering that cotton cloth and garments have been described as significant and valuable imports to Africa's eastern coast (Horton and Middleton 2000). Whether cotton seeds were used for oil or fiber production, the presence of cotton seeds in the archaeological record suggests that Pembans had access to locally-grown cotton.

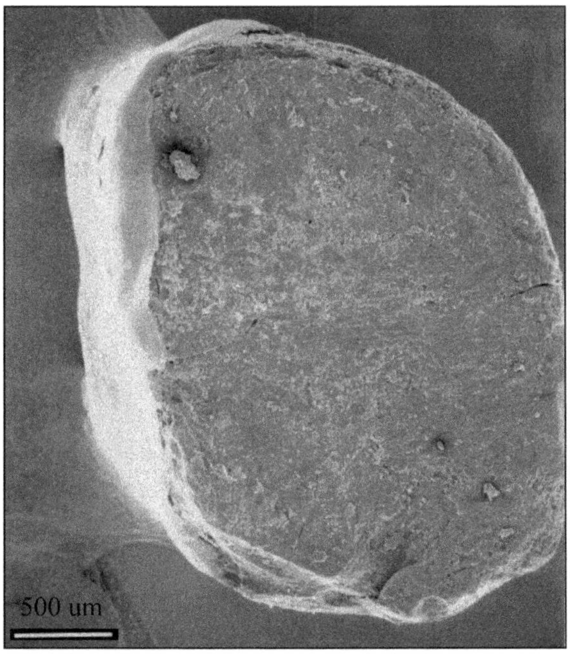

Figure 5.14. SEM image of *Vigna* sp. (possibly mung bean) cotyledon, medial view.

Interestingly, however, spindle whorls appear to be absent from Swahili sites before 1000 A.D.; none were recovered in deposits from Tumbe or Kimimba, and only one spindle whorl was recovered from contemporaneous deposits at Shanga (Horton 1996:396, Table 30). Another confounding factor in the identification of archaeological cotton is that baobab (*Adansonia digitata*) has similar seed anatomy to cotton, particularly when remaind are fragmented: both have a palisade layer and a "cap" structure (in cotton, this is called a funicular cap). Tellingly, baobab has recently been moved into the Malvaceae family. In light of this problem, we can perhaps be certain only that 2 cotton caps represent cotton at Tumbe; the remaining 9 non-cap fragments may in fact be baobab (Walshaw 2010). This identification problem is overcome at Chwaka (see Chapter 6) because of the finding of intact cotton seeds.

Figure 5.15. SEM image of cotton "cap". Scale is in micrometers

Other. One potential seed of the sesame plant (*Sesamum* sp., including the domesticate *S. orientale*) was recovered from Tumbe. Sesame, or *simsim* in many African and Asian languages, was domesticated in the Indus Valley of South Asia by 2000 B.C. (Bedigian 2004; Fuller 2003). While early researchers attributed sesame's origins to Africa based on the distribution of wild members of the genus, recent work on the morphology and genetics of this genus by Bedigian and others virtually rules out an African origin and favors an Asian domestication sequence (Bedigian 2003).

The author of the *Periplus* describes sesame oil as an Indian export to eastern Africa (Huntingford 1980), and the Swahili may well have obtained sesame oil through trade with southern Asian merchants. The finding of a seed, however, remains interesting, because oil pressing and carbonization are both expected to be destructive to the seed (Bedigian 2003). Wild sesame grows on Pemba, however, and has many uses such as food and medicine (Bedigian 2004), so in the absence of evidence confirming the Tumbe specimen as a domesticate, the possibility remains that it could be the wild taxa *S. angustifolium* (Koenders 1992:40).

Seeds & Chaff- Wild Plants

Carbonized seeds attributed to common weeds of cultivation were present but uncommon at Tumbe, with a ubiquity of 48% if the taxa are collapsed into one category. These include seven incidences of dock (*Rumex*) or another member of the Polygonaceae family, two seeds in the Euphorbiaceae (spurge) family, one seed of amaranth (*Amaranthus* sp.- but only partially charred), and three examples of an unidentified seed (UNID Seed "A") likely belonging to the Papaveraceae (poppy) family (see Figure 5.16). Amaranth is not only a common weed genus, but is an important vegetable plant in Tanzania, as indicated by nine species represented in a recent volume discussing the vegetable plants common in tropical Africa (Grubben and Denton 2004). *Rumex* and *Mollugo* are both common Old World weeds.

The rarity of small seeds is interesting in several regards. First, as discussed in Chapter 2, the presence of wild and weedy taxa has been associated with agricultural activities (weeding and harvesting) as well as the burning of dung for fuel (Miller and Smart 1984). It is important here to recall that the non-mechanized Swahili methods of harvesting rice and pearl millet by hand discriminate against the collection of weedy plants along with the harvest. Thus, the absence of weed seeds should not be interpreted as a lack of agricultural activities close to Tumbe. It is, however, unlikely that dung served as an important fuel source for the Swahili. Farmers today use animal dung to fertilize their fields, and this practice may have ancient antecedents. The scarcity of wild and weedy seeds is best explained by the deposition of weedy field plants away from the household, most likely in or near the fields themselves or as animal fodder.

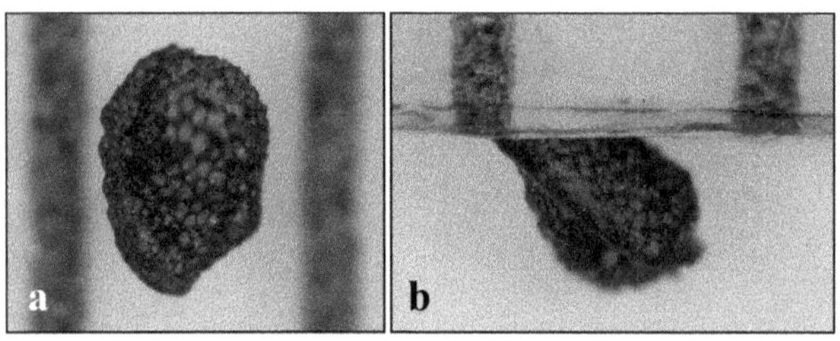

Figure 5.16. UNID Seed A, probably in the poppy (Papaveraceae) family: (a) side view; (b) view with ridge towards camera. Scale is in mm.

Non-Botanical Categories

Flotation samples may contain non-botanical items that are counted and weighed if larger than or equal to 2 mm in size. These finds typically mirror the artifact finds from the excavation screens, but occasionally novel items appear in the flotation samples, so they are included here. Non-fish faunal material was common in small amounts at Tumbe, and much more abundant than the few fish bones recovered. Non-fish animal bone fragments were found in 43% of Tumbe samples, while fish bones were only 8.7% ubiquitous. As will be discussed in Chapter 6, this stands in stark contrast to the pattern of animal remains in the Chwaka flotation samples, which were typically abundant in fish bone and lacking in mammal bone.

Small sherds of non-diagnostic pottery were found in thirteen samples (57%), while six samples (26%) contained daub fragments. Combining the pottery and daub finds into one general category of "ceramics" results in a ubiquity of 65%. Small pieces of stone or sedimentary concretions were present in 52% of samples. No beads or glass shards were found in the Tumbe flotation samples.

KIMIMBA (Period 1: 700-1000 A.D.), n=17

Seventeen archaeobotanical light fraction samples from Kimimba have been analyzed to date. Sample density (weight of flot/volume of sediment floated) on average is close to that observed at Tumbe: 5.11 g/liter. One flotation sample, Bag 205, had an extraordinarily high sample density (15.3 g/liter) due to a very heavy pan fraction – this can occur when the sand grain size is very small, and when the sample is overly agitated during flotation. Thus, the sample density for bag 205 is overstated. Two contexts appear to be wood charcoal concentrations and thus were truly the most archaeobotanically dense samples from Kimimba. The density of identifiable flot components averaged 1.75 g/liter, which is reduced to 0.94 g/liter if two wood-heavy samples are removed from consideration. Kimimba ubiquity data are provided in Figure 5.17, density data are shown in Figure 5.18. Weights and/or frequencies of all botanical items from Kimimba are presented in Table 5.4.

Preservation and Disturbance

All samples showed some signs of bioturbation – root or animal disturbance. Every sample contained uncarbonized plant parts, predominantly fine roots. Eighty-eight percent of samples contained uncarbonized seeds, and 76% had insect eggs and/or fungal growth. As with Tumbe, uncarbonized seeds were not counted towards the overall seed count, but were retained for comparative purposes. Kimimba site lies below fields of cassava and sweet potato, bordered by coconut trees, thus the presence of common agricultural weeds is not surprising.

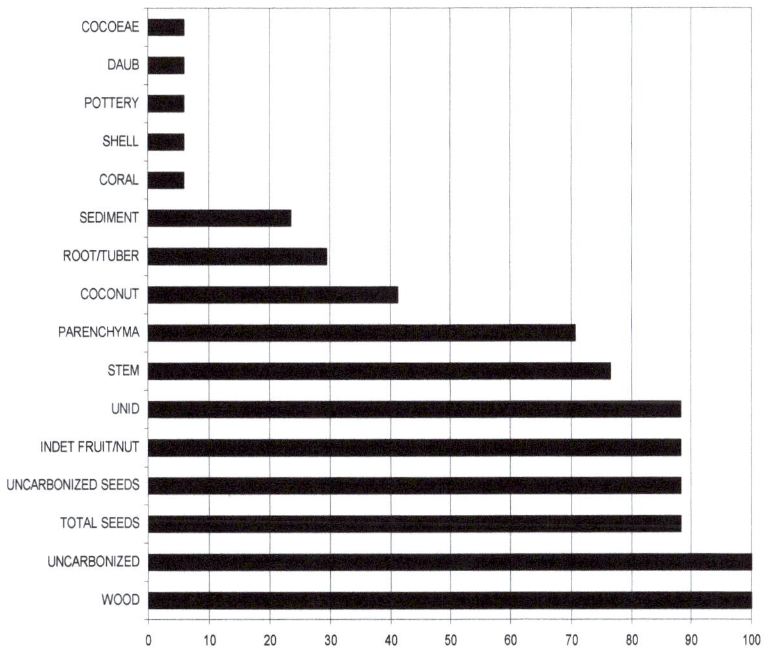

Figure 5.17. Percent frequency (ubiquity) of botanical categories among samples from Kimimba.

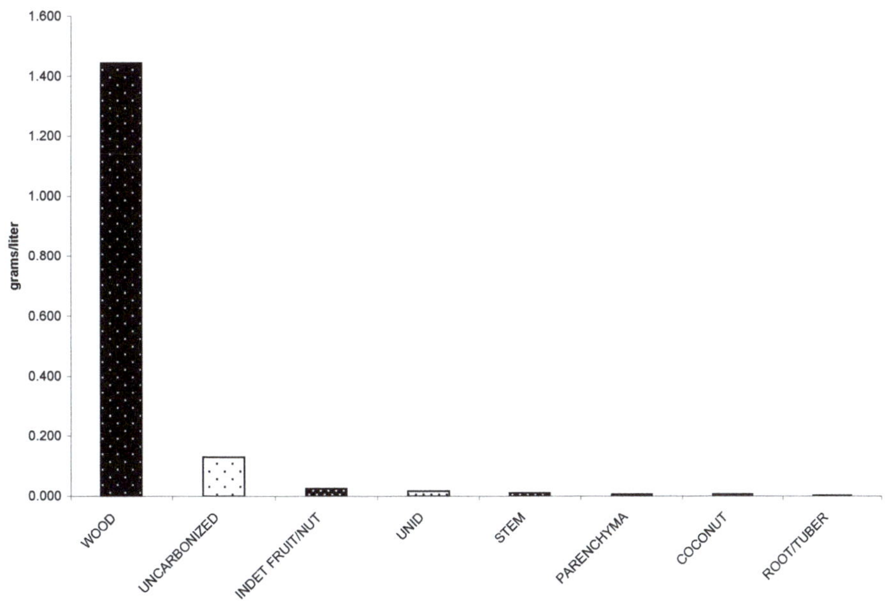

Figure 5.18. Density of botanical materials from Kimimba.

Table 5.4. Botanical representation among archaeological contexts (units) at Kimimba. D=density measure, U=Ubiquity measure.

	Kimimba Units												Site totals	
	3	**4**	7	**8**	**8b**	**8c**	**9**	**10b**	**11**	**13**	**14**	**15**		
Samples (n=)	3	3	1	2	1	1	1	1	1	1	1	1		
Volume (liters)	51	42	17	42	12	11	6	10	15	12	3	21	**17 samples 242 liters**	
													D(g/l)	U(%)
NON-SEED TAXA														
Wood (g)	48.9	52	.6	3.4	.9	23	43.1	8.0	2.0	23.2	12.0	129.6	**1.44**	**100**
Parenchyma (g)	1.07	.77	.01	.01	.01	.03	0	0	0	.02	0	.04	**<.01**	**70.6**
Stems (#)	21	150	1	0	0	5	8	0	1	3	2	22	**<.01**	**76.5**
Root/tuber (#)	1	33	3	0	0	0	0	0	0	0	0	0	**<.01**	**29.4**
Coconut (g)	.02	0.91	.01	0	0	.13	0	0	.01	0	0	.32	**<.01**	**41.0**
Fruit/nut (g)	.38	3.46	.01	.02	0.02	.01	7.96	0	.52	.13	0	1.94	**.02**	**88.2**
UNID (g)	1.07	0.51	.03	.02	0.01	0	.19	.02	.04	.01	.16	1.67	**.02**	**88.2**
Uncarbonized (g)	7.16	3.08	1.66	1.81	1.36	.98	.08	.07	3.74	1.46	.17	16.01	**.13**	**100**
SEED TAXA													**Total**	U(%)
Poaceae like Pearl Millet	0	0	0	0	0	0	0	1	0	0	0	0	**1**	**5.9**
Poaceae	0	0	0	0	0	0	0	0	1	0	0	0	**1**	**5.9**
***Macrytoloma* sp?**	0	1	0	0	0	0	0	0	0	0	0	0	**1**	**5.9**
UNID Fabaceae	4	4	0	0	0	0	0	0	0	0	0	0	**8**	**23.5**
Cotton	0	1	0	1	0	2	0	0	0	0	0	1	**5**	**23.5**
Fig	0	0	0	0	0	0	0	1	0	0	0	0	**1**	**5.9**
***Rumex* sp.**	0	1	0	0	1	0	0	0	0	0	0	0	**2**	**11.8**
***Rubus* sp.**	2	1	0	0	0	0	0	0	0	1	0	0	**4**	**17.7**
Cucurbitaceae	0	1	0	0	0	0	0	0	0	0	0	0	**1**	**5.9**
***Amaranthus* sp.**	1	0	0	0	0	0	0	0	0	0	0	0	**1**	**5.9**
UNID seed A	0	0	0	0	0	0	0	0	0	1	0	0	**1**	**5.9**
UNID seed B	3	0	0	0	0	0	0	0	0	1	0	0	**4**	**11.8**
UNID seed D	0	0	0	0	0	0	0	0	1	0	0	0	**1**	**5.9**
UNID seed E	0	0	0	0	0	1	0	0	0	0	0	0	**1**	**5.9**
UNID seeds	23	11	9	0	2	1	0	13	3	2	0	15	**79**	**76.5**
SEED TOTAL	**32**	**20**	**9**	**1**	**3**	**3**	**0**	**15**	**6**	**5**	**0**	**21**	**116**	**–**

Wood

Overall, wood density at Kimimba is 1.44 g/liter, but this is largely due to four wood-heavy samples with wood densities ranging from 2.1 to 7.18 g/liter. Interestingly, all these samples were taken from units located towards the outer limits of the site (Units 9, 13, 14, and 15) containing a low density of artifacts and architectural debris. The high concentration of wood in such contexts is puzzling, but could be attributed to the presence of a fire pit, or possibly the accidental burning of an incomplete wooden structure. A moderately high density of wood (1.23 g/liter) from bag 104 in unit 3 is most likely associated with an in-situ house-clearing episode, marked by intact charred logs sealed by midden deposits (LaViolette et al. 2003). Bag 104 is from the lowest of the midden fill, and likely represents the transitional zone between the house-clearing and the formation of the midden pit. Kimimba wood specimens are well preserved and preliminary attempts to identify wood taxa from three specimens showed them all to be different taxa, but none identifiable to genus (Alexa Höhn, personal communication 2003, Goethe University, Frankfurt). Future examination of the wood from Kimimba, and from other Pemban sites, is suggested, in conjunction with the collection of relevant botanical comparative material.

Storage Tissues

Parenchyma (definition provided above in Tumbe section) was present in 71% of samples, with one sample in particular containing a significant amount of parenchyma (see Figure 5.19 for high magnification image of parenchymal structure). The average parenchyma density was just below.01 g/liter, which is similar to Tumbe despite the fact that parenchyma is three times as common at Kimimba. Root/tuber specimens were found in five samples (29%), all of which come from units with high artifact frequencies, and thus are most likely associated with human activities as opposed to natural growth and incidental deposition. In particular, Unit 4 samples all contained root/tuber material. This low abundance of root/tuber is consistent with the findings from Tumbe at the same period, along with a mean density too low to be recorded in the density figure.

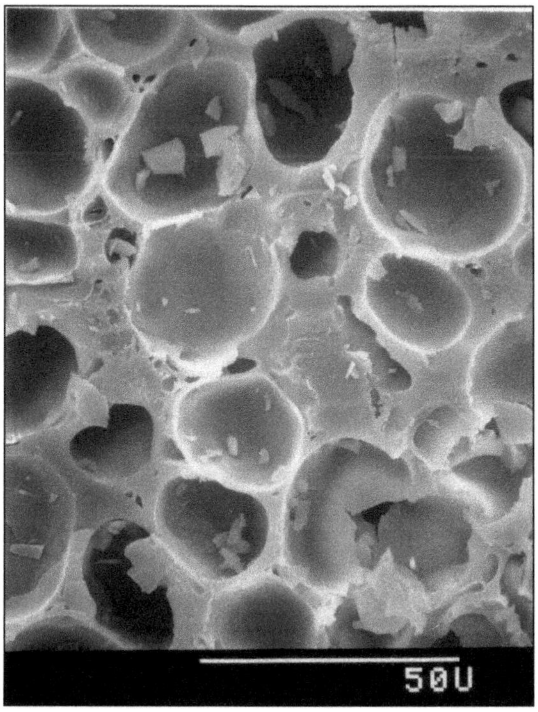

5.19 Archaeological parenchyma from Kimimba. Scale in μm.

Coconut (Cocos nucifera) *and Relatives*

Coconut's presence at Kimimba is relatively sparse compared to Tumbe: only seven samples (41.2%) have materials that could possibly be coconut endocarp, as opposed to 78% at Tumbe. Moreover, the Kimimba coconut is present at low density (<.01 g/liter), but this is comparable to samples at Tumbe from the same time period. Because no later occupations overlie the Kimimba contexts, the presence of coconut may not be attributed to contamination from younger deposits, as discussed above for Tumbe. Thus, the presence of charred coconut endocarp confirms its early presence and use by the Swahili. The relatively low abundance of coconut at Kimimba could be explained by any or all of several factors, including: a) infrequent use of coconut; b) low availability of coconut; c) re-use of coconut shells as containers or tools, not as fuel; d) the disposal of coconut debris away from the site; and/or e) different preservation conditions than those prevailing at Tumbe.

Other Fruit/Nut

Fruit and nut fragments are found more frequently than coconut endocarp at Kimimba, on par with Period 1 Tumbe fruit/nut frequencies. Over 88 % of samples contain whole or fragmented fruits, with an average density of .10 g/liter. This figure is reduced to .03 g/liter (same as the Tumbe Period 1 average) if we remove sample 108 (Unit 4) from consideration due to its relative abundance of well-preserved fruit/nut fragments (1.28 g/liter). Several of the assemblages contain good candidates for future identification efforts, such as the presence of partial pieces of a fig-like fruit (*Ficus* sp., see Figure 5.20), a whole palm fruit, one whole unidentified fruit with five carpels (see Figure 5.21) and several fragments which closely resemble one of the inner layers of the baobab fruit rind (*Adansonia digitata* - see figure 5.22).

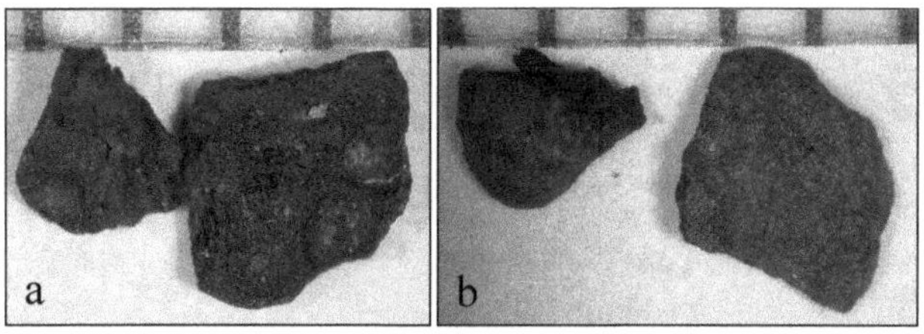

5.20. UNID fruit/nut from Kimimba resembling *Ficus* sp.: (a) inner surface; (b) outer surface. Scale in mm.

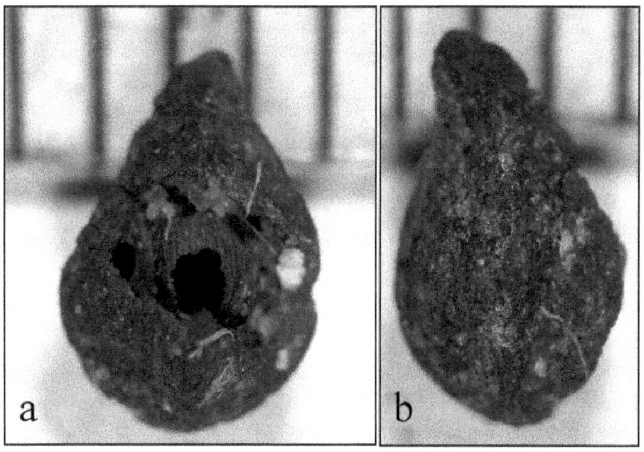

5.21. UNID whole five-sided fruit from Kimimba: (a) view showing seeds inside; (b) view of outer surface. Scale in mm.

Other Botanical

Botanical fragments that could not be attributed to any taxonomic or morphological unit (UNID BOT) were as common at Kimimba as reported for Tumbe, and were as common as fruit/nut fragments at 88 % frequency among the samples recorded. Unidentified botanical fragments were present, on average, at a density of .11 g/liter, which makes this the second densest category at Kimimba, far behind wood.

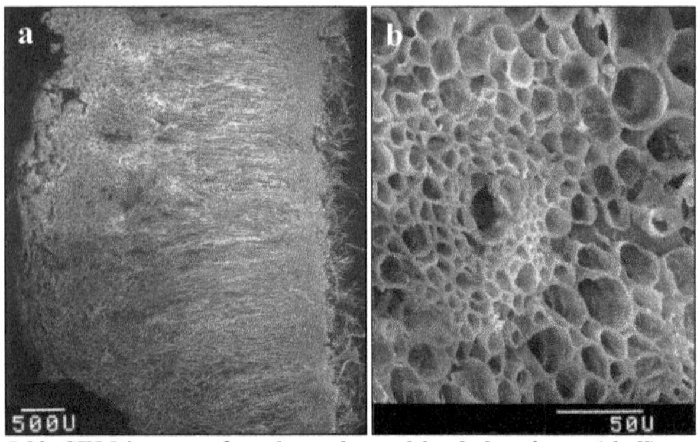

5.22. SEM images of modern charred baobab epicarp (shell), which compares favorably to some of the unidentified specimens from Kimimba; (a) cross section view; (b) detail of the inner surface. Scale is in μm.

Seeds and chaff – Domesticates

Pearl Millet. Of the thirty-six identifiable seeds recovered from Kimimba samples, only one possible pearl millet grain was found. This comes from Bag 193 (Unit 10b), from the context with abundant parenchymal fragments. No pearl millet spikelet bases were recovered, nor were any other recognizable domesticated grains or chaff, including finger millet, sorghum, and rice. Seed frequency data are illustrated in Figure 5.24.

Unidentified Grains and Chaff. One unidentifiable grass grain was found in Bag 182 (Unit 11). Stems were found in 70% of the Kimimba samples; no identifications were made, but the stems resemble grass straw and, as discussed above for Tumbe, may represent crop processing waste or house construction debris.

Cotton. Cotton fragments were found in 4 (24 %) of the samples, totaling an estimated five whole seeds. This abundance and frequency are similar to the cotton found at Tumbe. Spindle whorls were absent from the archaeological record at Kimimba, consistent with contemporaneous sites along the eastern African coast, which brings doubt to the identification of these fragments as cotton, when baobab is also expected to be present.

Fabaceae. Beans were the most frequently encountered seed taxon, being present in four samples for a minimum of eight beans total. Bag 113 from Unit 4 contained the richest Fabaceae assemblage: nine fragments of unidentified legume (SNE=3), and one
whole bean resembling the genus *Macrytoloma.* This context is one of the densest in artifacts and daub, and possibly represents an abandoned structure that was subsequently used as a refuse site for surrounding houses. The legumes are likely to be household waste from meal preparation. Beans are a common feature of the Swahili diet today, and are often included whole or mashed in maize and rice dishes (Hachten 1970; Ominde 1975).

Seeds & Chaff – Wild Plants

One possible fig seed (*Ficus* sp., likely wild) was recovered from Unit 10b (Bag 193), which is not unexpected considering the possible presence of fig fruit pieces in other parts of the site. Unit 4 produced one dock seed (*Rumex* sp.) and one unidentified seed of the melon family (Cucurbitaceae). Four seeds yet to be assigned a genus (UNID Seed "B", a large round seed resembling *Brassica* – see Figure 5.23) were found at Kimimba, three from Bag 104 in Unit 3. That same unit also contained the only amaranth seed found at Kimimba, as well as one of several examples UNID Seed "A", probably a member of the poppy family. Two other unidentified seed types ("D" and "E") were also found at Kimimba, shown in Figures 5.25 and 5.26. All four unidentified seed types are represented by good specimens, and future determinations are likely when adequate comparative specimens are available. The vast majority of samples contained abundant unidentified seeds (88%) or unidentifiable seed fragments (94%), and it is expected that future work will aid in the identification of this component of the Kimimba archaeobotanical record.

Non-Botanical Categories

Very few non-botanical artifacts were recovered from Kimimba. No bone of mammal or fish origin was recovered, and only one piece of shell was discovered. Five tiny sherds were found in the Unit 4 sample, one sample included daub fragments, and two contexts contained geofacts. The low abundance of material culture items could be due to natural or cultural factors. Kimimba sediments were very sandy, with large sand grains observable when conducting flotation; any sand adhering to artifacts would likely cause them to sink during flotation. Also, artifact frequency at Kimimba is lower on average than at Tumbe, thus one would expect less items overall.

5.23. UNID Seed B, resembling but not consistent with mustard (*Brassica*). Scale is in mm.

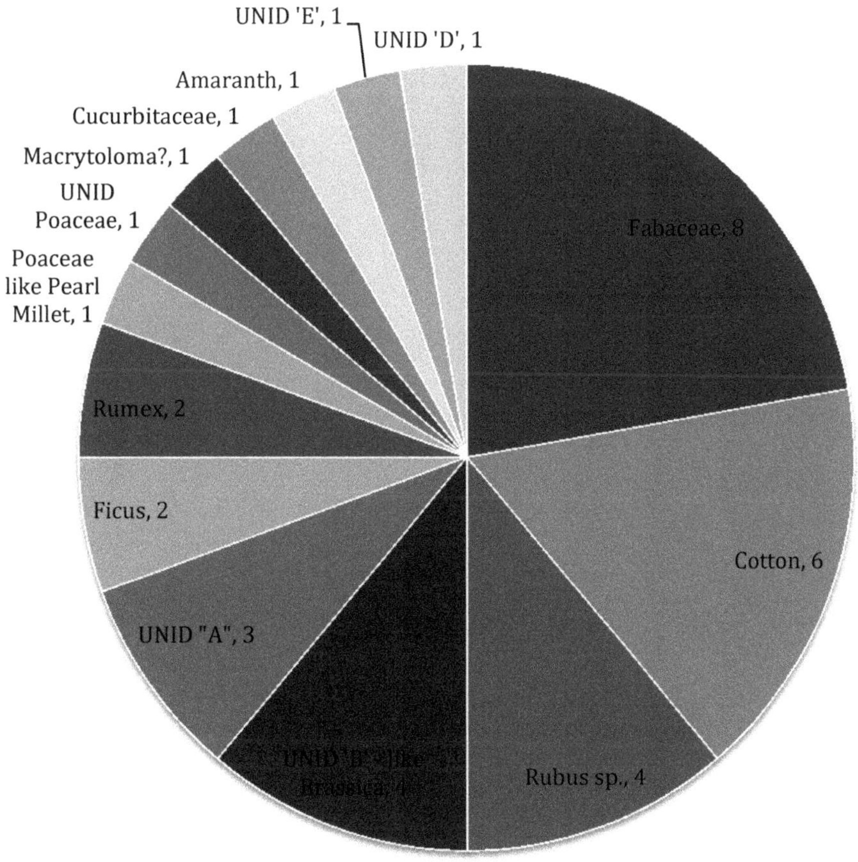

5.24. Frequency of seed taxa at Kimimba

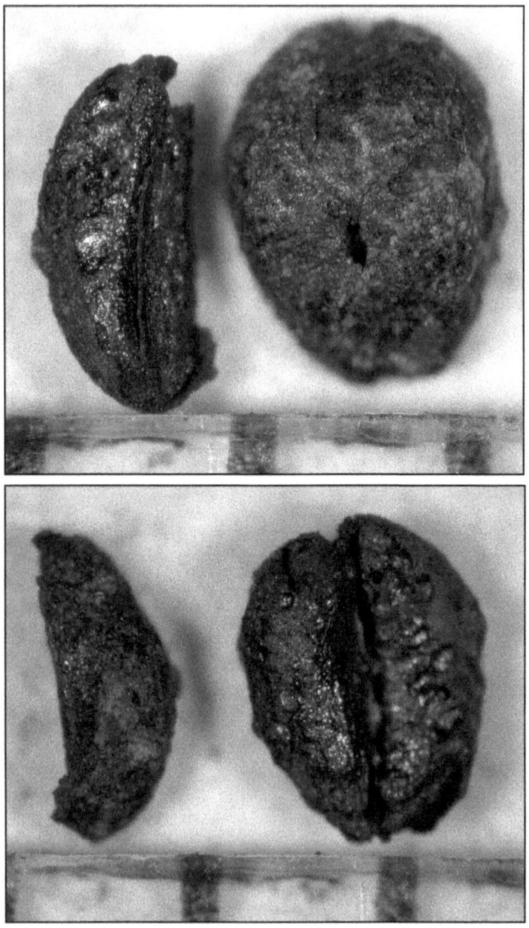

5.25. UNID Seed D, a four-part disseminule. Scale is in mm.

5.26. UNID Seed E, a capsular fruit. Scale is in mm.

Summary: Seventh to Tenth Century Archaeobotanical Patterns

This chapter presented the archaeobotanical data collected from the first millennium town of Tumbe and the contemporaneous village of Kimimba. This was done to facilitate analysis of their food production systems, particularly to determine whether connections exist, or if some production activities were focused in one site or the other. In Chapter 7, these data will then be compared with a later snapshot of Pemban food production, which is offered by Chwaka and Kaliwa and presented in Chapter 6.

Non-Seed Components

A close comparison of the plant remains from Tumbe and Kimimba reveals significant differences in access to food resources. Both contain an abundance of charred wood, which comprised just over 84% of the weight of sorted materials at both sites. Uncarbonized plant remains were the next most common feature at each site, indicating similar levels of post-depositional disturbance.

Coconut was the third most abundant category by density at Tumbe, while this position was occupied by fruit/nut remains at Kimimba, and this represents the first major differentiation between the two sites. In fact, coconut was much more sparsely represented at Kimimba, and may represent a differential access to coconuts or, less likely, a preference for wood as fuel. Similarly, the fruit/nut category is much better represented at Kimimba, almost ten times as important when observed through density measurements. While the lack of positive identification of fruit/nut materials prevents me from making detailed comments about the relative abundance of this category, it is most likely that they represent an indigenous food resource, in contrast to the probably introduced coconut. This suggests that the residents of Kimimba may have had more limited access to exotic or introduced foods than those of Tumbe.

Only minor disparity between the sites can be seen in the relative abundance of parenchyma and root/tuber fragments. Parenchyma is relatively rare at Tumbe, as are roots and tubers. At Kimimba, however, parenchyma is slightly more common than coconut, and several good examples of root/tuber organs were found. As with the fruit/nut category, the lack of strong identifications for these items limits my ability to interpret the significance of this discrepancy. This may suggest that Kimimba residents were more reliant on one or more of the starchy foods (yam, banana, and taro) and small bulbs/roots/tubers than the residents of Tumbe.

Another source of differentiation can be found in the relative absence of stems at Tumbe compared with a more significant presence at Kimimba. While stem identification to taxon remains to be accomplished through phytolith analysis, stems were most likely contributed from either architectural sources (thatching, daub temper, patching) or from processing of grass crops, including grains. Tumbe houses were arguably constructed with the same methods and materials as Kimimba's, so using the architectural hypothesis, it is somewhat striking that Tumbe lacks stem remains. However, the main grain found at Tumbe is pearl millet, which has a much thicker and distinctive stem. It remains a possibility, therefore, that some plants (for example rice, see below) were processed at Kimimba, but not at Tumbe; and it also remains possible that some plants were processed at Kimimba for consumption at another site, such as Tumbe. This leads us to a discussion of the grains and other seeds present at Tumbe and Kimimba

Seeds and Chaff

Seventeen types of seeds were recovered from Tumbe, with pearl millet and possible pearl millet grains dominant at 75% of seeds by abundance. In fact, all seeds other than pearl millet and unidentified seeds comprise just less than 14% of the total amount of seeds at Tumbe. Thus, while at first glance Tumbe appears to enjoy a diverse array of seeds, in reality pearl millet is dominant. After pearl millet, cotton is the next most abundant seed type, followed closely by dock (*Rumex* sp.) and unidentified members of the bean family. The following seed types have frequencies of less than 2%: rice; UNID millet; a member of the wheat tribe; sorghum; *Vigna* sp.; UNID type "A"; UNID Poaceae; sesame; and members of the dock, spurge, and purselane families.

Fourteen seed types were found in samples from Kimimba, but none overwhelmingly dominant as in Tumbe. In fact, the Kimimba pattern closely resembles the non-pearl millet portion of Tumbe's seed record, except that no positively identified grains are present. Also, Kimimba shows evidence of a different food plant repertoire: in place of grains are beans and fruits (*Rubus*, melon, fig), with common weed taxa present, including amaranth (possible eaten a leafy green, or present as a weed) and dock. Kimimba also shows evidence for three well preserved but unidentified seed types not observed at Tumbe, which may represent local food resources because imported foodstuffs are typically better known botanically than the local flora of Pemba.

Like the pattern observed for other domesticates, Tumbe and Kimimba could not be more different where chaff is concerned. No chaff fragments from any grain were reported from Kimimba, while both pearl millet and rice spikelet bases were present at Tumbe. In fact, rice spikelets far outnumber rice grains and account for most of the ubiquity measure of 30% for this taxon. This can reasonably be interpreted one of several ways. One possible meaning of this pattern is that Kimimba residents did not participate in grain agriculture in any serious way, while Tumbe relied heavily on pearl millet cultivation, and, to a lesser extent, rice cultivation. A different argument altogether could be made, however. According to some archaeobotanists, sites at which grain is cultivated, processed, and shipped off for consumption elsewhere (the "producer" sites mentioned in Chapter 2), will show evidence for only the early stages of crop processing (Hillman 1984; Jones 1984). This would be almost invisible archaeobotanically with crops such as pearl millet and rice, both of which can be shipped while the grains remain on the head. This is particularly true of hand-harvested crops, as no weed seeds accumulate using this traditional non-mechanized method likely used by the ancient Swahili. Still a third explanation exists for this pattern of crop plants. Kimimba may represent a seasonally or culturally different type of occupation from Tumbe. I will return to this hypothesis in more detail in Chapter 8.

In summary, the more populous town of Tumbe seems to have had access to introduced plants, and relied more on domesticates such as pearl millet, than the contemporaneous rural village of Kimimba. Instead, Kimimba residents appear to have expoited a wider range of local food resources, including fruits and starchy plant organs, compared to the Tumbe population. If the people of Kimimba grew crops such as pearl millet and rice, there is scant evidence to prove that they either ate it or produced it for consumption at Tumbe. The archaeobotanical evidence constructs an image of two relatively independent communities where food production is concerned. Following the presentation and discussion in Chapter 6 of archaeobotanical data from the second millennium sites of Chwaka, Kaliwa, and Mazrui Tumbe, I will return to a discussion of crop production in Chapter Seven.

CHAPTER 6: ARCHAEOBOTANICAL RESULTS FROM SECOND MILLENNIUM A.D. PEMBA: CHWAKA, KALIWA, AND MAZRUI TUMBE

Introduction

Here I report the 2nd millennium A.D. archaeobotanical data, some of which has been previously reported in abbreviated form (Walshaw 2010). This includes all levels from the 11th-15th century town of Chwaka, all levels from Kaliwa, a 14th-15th century village, and the upper levels of Tumbe associated with Mazrui habitation of the site in the 19th century. Chwaka and Kaliwa are presented in the same chapter to permit comparison between town and village toward the end of the occupation of Chwaka. The Mazrui levels at Tumbe are presented here to reduce confusing the 19th century occupation with the 7th-9th century town described in Chapter 5. As noted in the previous chapter presenting the 1st millennium A.D. results at Tumbe and Kimimba, composite samples were generally 10 or 20 L in size, while feature samples were as large as possible, but often measured less than 5 L. To control for size differences between samples, materials greater than or equal to 2 mm are reported as densities (g/liter), as well as ubiquities (% of samples with taxon present).

CHWAKA: Eleventh to Sixteenth Centuries A.D. (Periods 2-4), n= 34

Six feature samples (2-10 L) and 28 composite samples (8-20 L) contributed to the data set presented here. On average, Chwaka samples were very rich in archaeobotanical materials, but some samples were extraordinarily rich in particular taxa, such as rice or cotton. The mean flot density for the Chwaka samples analyzed to date is 7.17 g/liter – somewhat higher than that from Tumbe and Kimimba. Bag 1281 (Chwaka Op-2B[1]) contained an enormous amount of charred rice, explaining its high flot density of 19.36 g/liter. The average flot density for Chwaka Operation 2 samples (n=16) came to 9.86 g/liter, while the remaining units and Op-1 (n=18) averaged only 4.78 g of flot per liter of sediment floated, making Op-2 twice as archaeobotanically rich as the other units combined. Much of the high density of archaeobotanical material was restricted to the lower levels of Chwaka Op-2, dated to approximately Periods 2 and 3 (eleventh to thirteenth centuries A.D.). Looking at sample richness using the measure of identifiable flot density, all samples averaged 3.72 g/liter, while Op-2 reached an average of 5.99 g/liter, largely due to a probable hearth feature dense in wood, but also due to the density of rice in the lower levels.

Preservation and Disturbance

Chwaka samples contained very well preserved specimens, with little uncarbonized materials or other evidence of disturbance. All samples contained uncarbonized plant remains (100% ubiquity; see Table 6.1, Figure 6.1) but this averaged only .08 g/liter of sediment floated (compare with .42 g/liter at Tumbe and .13 g/liter at Kimimba; see Figure 6.2 for Chwaka density data). Almost all (91%) samples contained uncarbonized seeds with a mean of almost 22 seeds per sample, and an average of almost 2 seeds per liter across the entire site. Uncarbonized seeds at Tumbe were common weeds associated with crops, which is to be expected considering Chwaka's location among cassava fields. Finally, only 38% of samples indicated bioturbation through insect activity, as evidenced by insect eggs, dung, and fungus fragments.

Wood

As discussed in Chapter 5 for Tumbe and Kimimba, wood is present in all samples and comprises the most abundant category by weight. The mean wood density at Chwaka is influenced by a few samples with abundant wood fragments, likely hearth features. Taken together, the samples have a mean wood density of 1.76 g/liter, comparable to the mean of all samples at Kimimba but smaller by half compared with

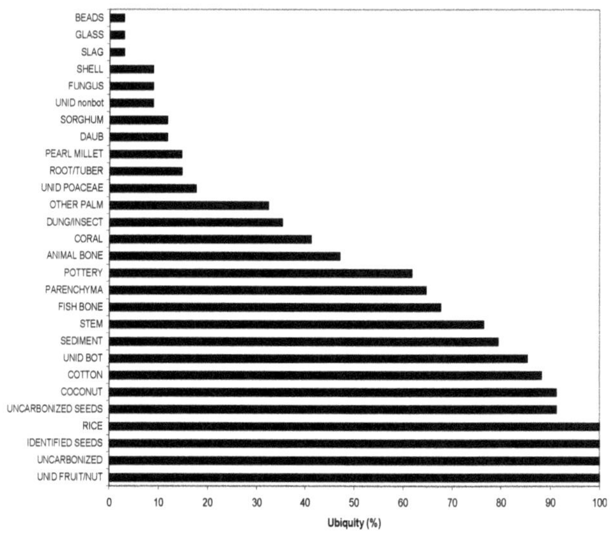

Figure 6.1. Ubiquity of materials recovered by flotation at Chwaka, including all sorted materials and the most common seed types.

Tumbe. One small feature sample (.5 L) contained wood at a density of 34.4 g/liter (Op-2C F-188), and its location near the wall of a house is consistent with a cooking fire. If we remove the four largest outliers, the Chwaka wood density decreases to .47 g/liter, which is comparable to Kimimba's wood density with the densest samples removed.

Storage tissues

Parenchyma was identified in 64.7% of the Chwaka samples, on par with Kimimba but much more ubiquitous than among Tumbe samples. Readers should note that distinctions between parenchyma and undetermined fruit/nut at Chwaka were easily blurred,

[1] Operations represent extensive excavations and are named by 2m x 2m sections within each operation. Op-2B was adjacent to Op-2C, and will be discussed several times in this report due to the significant finds therein.

Table 6.1. Non-seed archaeobotanical results from Chwaka, by archaeological context (Unit or Operation).

Unit	Samples (n=)	Volume floated (liters)	Wood (g)	Parenchyma (g)	Root/Tuber (#)	Stems (#)	Coconut (g)	Fruit/Nut (g)	UNID (g)	Uncarbonized (g)
11	3	66	33.61	.15	1	76	7.69	25.32	4.82	1.22
12	1	20	6.99	.21	0	13	2.07	9.43	.13	.89
19	1	10	4.50	.17	0	27	1.76	5.59	0	.07
22	1	8	7.32	0	1	70	2.27	15.92	.01	.35
26	1	12	.38	.05	2	2	.22	.25	.14	1.94
27	2	25	19.44	.40	0	84	4.81	8.63	1.19	2.10
38	1	2	.63	0	0	0	0	.02	.07	.26
39	1	12	.78	.04	0	0	1.13	.61	.08	3.50
OP-1A	1	12	1.55	0	0	22	.33	3.54	.01	.07
OP-1B	2	19	16.15	.59	0	9	2.63	3.79	.24	1.77
OP-1C	2	17	7.17	.05	1	2	1.31	1.82	.50	.39
OP-1F	1	8	1.02	.08	0	28	.5	1.68	.07	.06
OP-2A	1	9	3.25	0	0	2	0	.03	.04	4.15
OP-2B	9	73	73.65	.28	0	135	2.73	1.64	.87	13.32
OP-2C	3	14.5	29.1	.16	0	4	1.32	5.13	.48	1.71
OP-2D	2	22	5.19	.33	1	12	.96	5.93	.33	7.23
OP-2E	1	8	.49	.06	0	1	.01	.05	.02	5.91
OP-5	1	14	13.3	.13	0	5	3.0	7.19	.12	3.56
Site Totals (n= 34; Volume= 342 liters)										
Density (g/liter)			**1.76**	**.02**	**<.01**	**<.01**	**.10**	**.28**	**.02**	**.15**
Ubiquity (%)			**100.0**	**64.7**	**14.7**	**76.0**	**91.2**	**100.0**	**85.3**	**100.0**

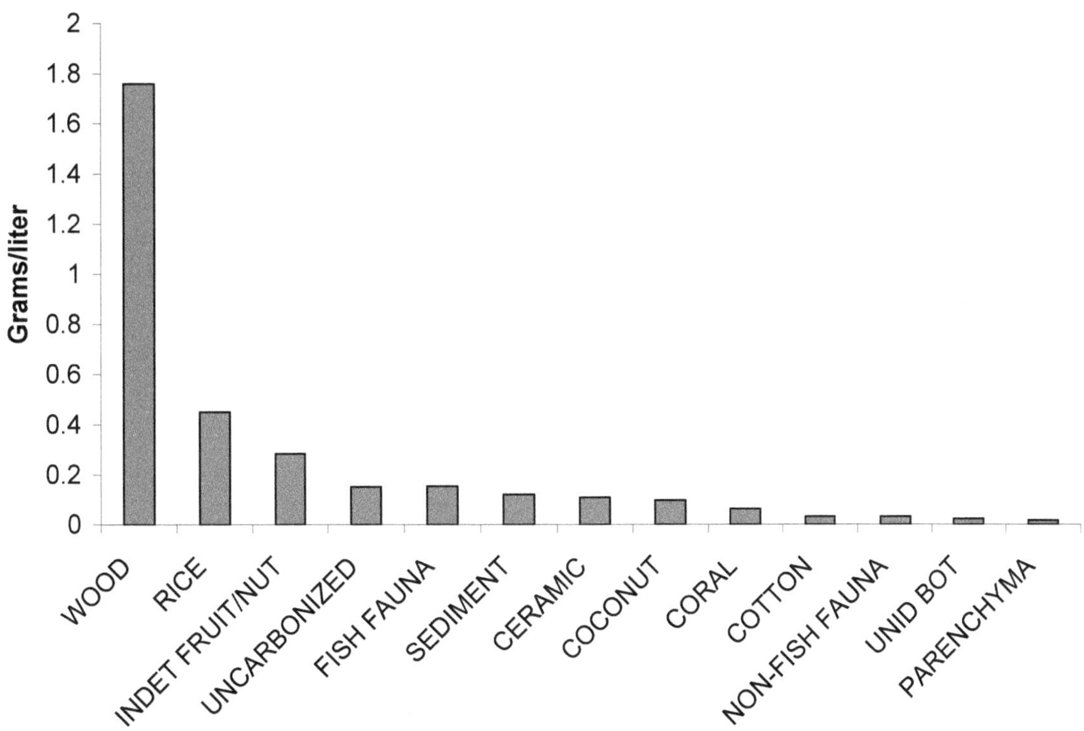

Figure 6.2. Density of materials recovered by flotation at Chwaka.

and future work on the taxonomic identification of both fruit taxa and root crops should aid in resolving this problem. For example, baobab rind has several layers, and when missing the outermost portion of the rind, the inner layers contain undifferentiated cells and thus may get labeled as parenchymal tissue (this can be seen in Figure 5.21 located in the previous chapter). Only six root or tuber fragments were observed in any of the Chwaka samples.

Coconut (Cocos nucifera) *and Relatives*
Like samples from Tumbe, coconut endocarp ("shell") fragments are common elements of the Chwaka archaeobotanical record. Over 91 % of samples contained coconut, with an average of 37.7 fragments per sample, and a mean density of .1 g/liter. This density of coconut is almost 10 times that at Tumbe, and over 10 times Kimimba's coconut density. The increased abundance of charred coconut suggests that endocarp fragments may have been used as fuel, in cooking fires, or in coconut production, particularly in the drying of coconut meat to make storable copra. This is consistent with the lower values for wood densities, as coconut may have gained in use as a fuel source. Coconut tree products have many uses, and the husks, shells, "meat" and fluid of coconut fruits are valuable as materials for fiber production, container and tool production, and, of course, as food. The intensification of any or all of these products may have led to increased coconut processing and subsequent disposal of coconut endocarp.

In addition to coconut, a few fragments of unidentified palm (possibly oil palm, *Elaeis guineensis*, and/or other unidentified members of the coconut tribe) were recovered from 11 (32.3%) samples. Oil palm fragments are best identified using a combination of morphological features, including the presence of an "eye" (pore) in the surface, linear stripes along the outer surface, and a curvature to the endocarp not seen in its larger relative, the coconut (A. Catherine D'Andrea, pers.comm.). Oil palm fruits are obovate and contain inner compartments which were not observed in any of the Chwaka unidentified palm fragments. Under a scanning electron microscope, spherical phytoliths typical of the palm family can be seen on some specimens of unidentified palm shell, and are even visible under high-powered light microscopy (see Figure 6.3). Non-coconut Cocoeae and palm fragments were not found in sufficient density to include them in the Chwaka archaeobotanical density figure but were included in the total fruit/nut density measure.

Other Fruit/Nut
Unidentified whole and fragmented fruits and nuts are also a common category of botanical finds at Chwaka. At one stage in the analysis, I had attempted to separate fruit and nut fragments by morphology, but this was soon ended when the categories overstepped the limits

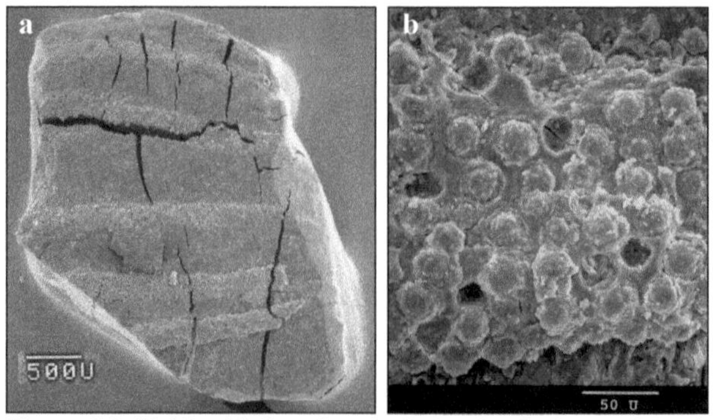

Figure 6.3. Unidentified archaeological palm fruit, possibly oil palm (*Elaeis guineensis*), imaged through scanning electron microscopy: (a) view of outer endocarp surface; (b) close-up of phytoliths on fibers seen on outer surface. Scale is in μm.

of my comparative collection. As previously indicated, I hope to study the Pemban fruits further when I collect adequate samples for comparison. For the purposes of the present discussion, therefore, the only fruit/nut I discuss taxonomically is coconut, provided above. The remaining fragments were present in all samples observed, exceeding even coconut in ubiquity. The mean density of fragments was .28 g/liter, almost three times that reported in Chapter 5 for Tumbe and Kimimba. Fruit/nut density is also almost three times that of coconut, although readers should note that some possible fragments of coconut are put into the unidentified fruit/nut category because they do not conform to the minimum morphological requirements of the coconut category. Moreover, coconut husk material has not yet been characterized for botanical identification at the level of small fragments, so large portions of the category may in fact constitute unrecognized coconut mesocarp (husk). As discussed in Chapter 5, the fruit and nut fragments could come from fig trees (*Ficus sp.*), of which there are 14 wild species and four cultivated species recognized on Pemba (Koenders 1992:32-33), or the baobab (*Adansonia digitata*), the locally recognized *mbono* (*Jatropha* sp. or possibly the outer casing of the castor bean, *Ricinus communis*, both Euphorbiaceae), or any number of local or introduced fruits.

Taken together, all coconut, palms, and unidentified fruits comprise an impressive portion of the Chwaka archaeobotanical assemblage. These fruits were found in all samples, and averaged .38 g/liter, approximately four times that of Tumbe and Kimimba.

UNID Botanical Material
Charred botanical materials that could not satisfactorily be assigned to any of the above categories, or to a seed taxon, were placed in this class. Eight-five percent of Chwaka samples contained some unidentified materials, averaging .02 g/liter.

This density is slightly above that of Tumbe and Kimimba, but not unexpected considering the sheer volume of finds in the Chwaka flotation samples.

Seeds and Chaff – Domesticates

Rice (Oryza *sp.*) Rice dominates the Chwaka archaeobotanical record, not only in the seed fraction (see Table 6.2, Figure 6.4), but sometimes an entire sample, as can be seen in Figure 6.2, where rice is second only to wood in density. Rice is 100% ubiquitous among the samples, with rice rachillae present in every sample and grains present in 88.2% of samples. A total of 15,146 rice grains (see Figure 6.5) were counted or estimated, with 99% coming from

Table 6.2. Seed and chaff frequencies among Chwaka contexts (Unit, Operation).

Seed Taxa (grains unless stated otherwise)	Unit 11	Unit 12	Unit 19	Unit 22	Unit 26	Unit 27	Unit 38	Unit 39	Op-1A	Op-1B	Op-1C	Op-1F	Op-2A	Op-2B	Op-2C	Op-2D	Op-2F	Op-5	TOTAL	Ubiquity (%)
Pearl millet					1					4	4			1					**10**	15
Finger millet														1					**1**	3
Sorghum														3				2	**5**	9
Sorghum chaff				1															**1**	3
UNID Millet	1								1	7				3					**12**	12
UNID Millet chaff		2								1									**3**	6
Triticeae					1														**1**	3
Rice	8	5		3	5	3	1	1	1	7	1		24	15040	27	4	13	3	**15146**	88
Rice chaff	63	84	6	18	38	13	1	23	47	119	35	15	8	4218	609	52	2	126	**5477**	100
Digitaria					1														**1**	3
UNID Poaceae	1									2				2					**5**	9
UNID chaff Poaceae					6														**6**	3
Vigna sp.										1					1				**2**	6
Mung bean	1							1						1		1			**4**	12
Pea	1					1								40				1	**43**	18
Fabaceae like Pea														31					**31**	3
Fabaceae	1			1		2				1				1	1				7	18
Cotton	32	10	6	20	2	32	1		9	1	5	5		16	406	9	1	36	**591**	88
Fig	2																		**2**	3
Poppy																		1	**1**	3
Dock	5		1			1													7	9
Dock family					1	1			1										**3**	9
Buttercup family					1														**1**	3
Pink family	1		3		1														**5**	9
Purselane family								1											**1**	3
Goose-foot family														1					**1**	3
UNID seeds	1	3			2	5		1	3	3		1		24	1	4		2	**50**	59

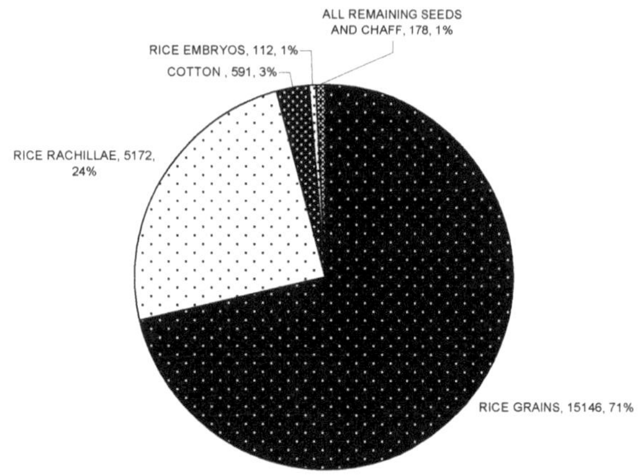

Figure 6.4. Chwaka seeds and chaff by frequency.

Chwaka Op-2B in contexts dated to the early middle occupation of this set of superimposed house structures (levels 8-5, 3). For example, Bag 1281 yielding 60.65 g of rice grains (estimated to be 6000 grains[2]). A total of 5172 rachillae (see Figure 6.6) were estimated, with 81% of these coming from Op-2. Bag 1281 at Op-2 produced .35 g of rachillae, estimated to be 1750 rachillae, and bag 1611 (also Op-2B) contained 352 rachillae (.07g).

The Chwaka samples also contained other rice chaff elements, including the palea and lemma (rice husk pieces), and isolated rice embryos (called the "germ"), a by-product of intensive rice cleaning. Five (15%) of Chwaka samples showed evidence for lemma and palea fragments, although several of the grains retained some of their chaff, so this measure may be underestimated. A total of 112 rice embryos freed from the grain were found in 16 (47%) of the samples. Thus, the Chwaka samples suggest rice cleaning was practiced in addition to rice cooking and eating. Removal of the rachilla suggests that Chwaka residents were processing rice for storage and/or consumption, and were likely obtaining threshed rice straight from their or others' fields.

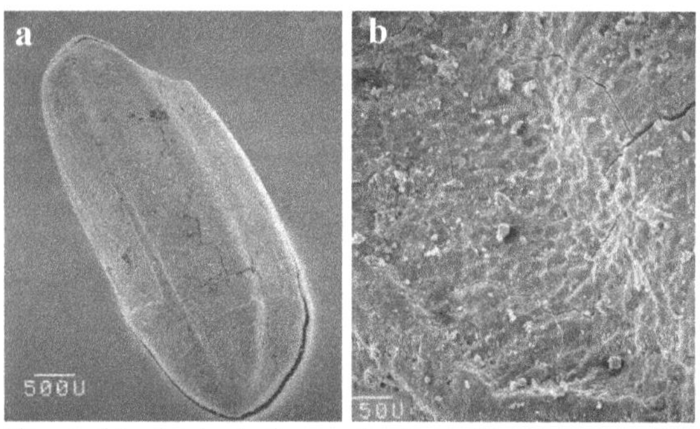

Figure 6.5. Ancient rice grain from Chwaka imaged using scanning electron microscopy; (a) whole grain; (b) close-up of bran layer adhering to seed. Scales are in μm.

6.6. Archaeological rice rachillae from Chwaka OP-2B: (a) viewed under SEM (scale in μm); (b) viewed under light microscope (scale in mm).

The most prominent crop in Asia, rice is easily identifiable and frequently recovered from archaeological contexts. Indeed, the prominent Chinese archaeobotanist Zhao Zhijun (Zhao 1998) estimates that the "emergence [of rice] as a domesticated subsistence plant drives much of the interest in research in archaeology in South and East Asia." During several millennia of cultivation, Asian rice became diversified into thousands of varieties, which have been grouped into three subspecies: japonica, indica, and javanica forms (T.-T. Chang 1976; Zohary and Hopf 2000:90). The indica plant is largely tropical, and as such is characterized by adaptation to a short day, and by a large variability in growth height, panicle size, and tillering capability. Japonica group cultivars tend to be subtropical to temperate plants that are not as sensitive to day length, but attain only modest heights and bear relatively small panicles. Among Asian rices, indica types are distinguished by a long, thin grain whose length is more than three times its width (Rehm and Espig 1991:11). Japonica type grains are round to oval, with depths exceeding the width by a small margin. The Javanica variety, originally from Indonesia, bears a broader grain than the other two subspecies (Zohary and Hopf 2000:90). Colloquially, the indica rices are categorized as "long-

[2] Seed number in this case was estimated using the weight of 100 whole carbonized rice grains from the sample. This method of estimation provides the most conservative value; a more accurate seed number estimate would be achieved by counting all the rice embryos, similar to the Minimum Number of Individuals measure used by zooarchaeologists.

grained", and the japonica and javanica are considered "short-grained", with the distinction having more to do with width and breadth dimensions than with absolute length.

Determining the ancestry of cultivated Asian rice is difficult because the domesticated forms are interfertile with many wild and weedy rices, the modern distributions of which are sufficiently large that a discrete area of likely domestication is not discernable (Zohary and Hopf 2000:90). Domestic rices most likely developed from wild forms similar to annual *O. nivara* and the perennial *O. rufipogon* (Chang 1976), although isozyme studies indicate that the latter is the most direct ancestor (Wang 1994 in Crawford and Shen 1998). Morphological change arose as a consequence of selection for the following traits : larger plants with larger grain size, synchronous tillering of non-shattering heads, shorter seed dormancy, and a weaker photoperiod response (Chang 1976). The earliest evidence for cultivated rice comes from several archaeological sites in the middle and lower Yangtze River Valley (China), from levels dating to the mid seventh millennium B.C. (Crawford and Shen 1998). Outside China, the earliest indication of rice cultivation is known from India later in the third millennium B.C., with intensification in production evident by the time of the Indus civilization at 2000 B.C. (Fuller 2000; Zohary and Hopf 2000:91).

Compared with other cereals, rice has a very short embryo, no furrow, and has several ridges running longitudinally along the grain. The rice grain is narrow (laterally compressed) but long and deep, giving it an oval to oblong shape when viewed from the side. A distinctive checkerboard pattern is often visible on the grain; this surface ornamentation, combined with evidence of an oval cross section and longitudinal ridging allows for the confident identification of even fragments of rice grains. Rice rachillae are distinctive from other millet types and are small but robust, evidenced by their 100% ubiquity among Chwaka samples.

Rice has been discussed by Swahili scholars as an Asian import; few have questioned whether rice on the eastern African coast could be the African domesticate, *O. glaberrima* grown widely in West Africa today. This is largely because rice appeared ethnohistorically to rise in importance on the coast concomitant with other Asian influences in material and culture, that is, after A.D. 1000. This has gained recent support with archaeobotanical finds from northeastern African ports. Asian rice attributed to Indian trade has been found at Roman sites in Egypt (Berenike, Quseir al-Qadim) dated by association to the first and second centuries A.D. (van der Veen 2003a; Wendrich et al. 2003). Hoffman (in Wright 1984:49) reports Asian rice in a ninth to tenth century A.D. archaeological sequence on the Comoros Islands, noting that "[t]he grains are relatively long, similar to the traditional varieties now grown on the Comores." Johnson (in Wright 1992) found that rice continued to be used by the thirteenth century on the Comoros. Claims to an Asian origin for eastern African rice, however logical, require scientific verification such as that provided here through metric analysis.

Measurements conducted on rice grains from Chwaka show that while the rice dimensions show a wider variation than found among modern cultivars today, the rice at Chwaka is most likely short-grained (see Figures 6.7 and 6.8). Short-grained rice is an over-arching category that transcends species: both Asian (*O. sativa*) and African (*O. glaberrima*) rice species contain short-grained types. Within Asian rices, the indica type is long grained, while the japonica and javanica types are short grained. As seen in Figure 6.6, the Chwaka rice overlaps with both of the short grained Asian types, but is particularly close to the temperate adapted rice, japonica. This is particularly clear when comparing mean width and breadth dimensions for each category (see Figure 6.8), where the Chwaka rice plots closest to Asian japonica rice. Interestingly, Chwaka rice also overlaps in its smaller range with *O. rufipogon*, the Asian wild rice thought to be the ancestor of the domestic Asian rices.

As discussed with pearl millet in Chapter 5, the finding of wild-sized rice at Chwaka can mean any of several things. First, it probably reflects recent changes in modern cultivar morphology and a narrowing of the genetic base of these highly studied and controlled crops. Thus, ancient domesticates may be smaller and/or more diverse in size than their modern counterparts. Second, it may signal that both wild and domestic morphotypes were introduced to eastern Africa, a likely scenario that would increase the genetic robusticity of crops and increase yields. Yet a third possibility remains that the archaeobotanical population is skewed by depositional or preservation biases. For example, smaller grains are less often retrieved when lost during processing or cooking activities compared with larger grains. I think this last scenario is the least likely considering that the metric sample came from a large rice concentration in a possible storage feature located in a house structure, and as such would represent a primary, not accidental, assemblage of grains.

African millets. African millets are rare at Chwaka relative to rice and also compared to the 43% ubiquity of pearl millet at Tumbe. Pearl millet grains were recovered from five samples (14.7%), totaling 10 grains throughout the entire site. No pearl millet spikelet bases were recovered. Three samples contain millet seeds that could be attributed to pearl millet, raising the total possible pearl millet to 20 grains at a ubiquity measure of 20.6%. Thus, whereas pearl millet constituted the dominant seed category at Tumbe by a significant margin, at Chwaka pearl millet is only the fourth most common identifiable seed taxon as measured by presence among the 34 samples (see Figure 6.9). Interestingly, three of the six Op-1 contexts sampled contained pearl millet, all from the upper levels (1-3) that are associated with the later

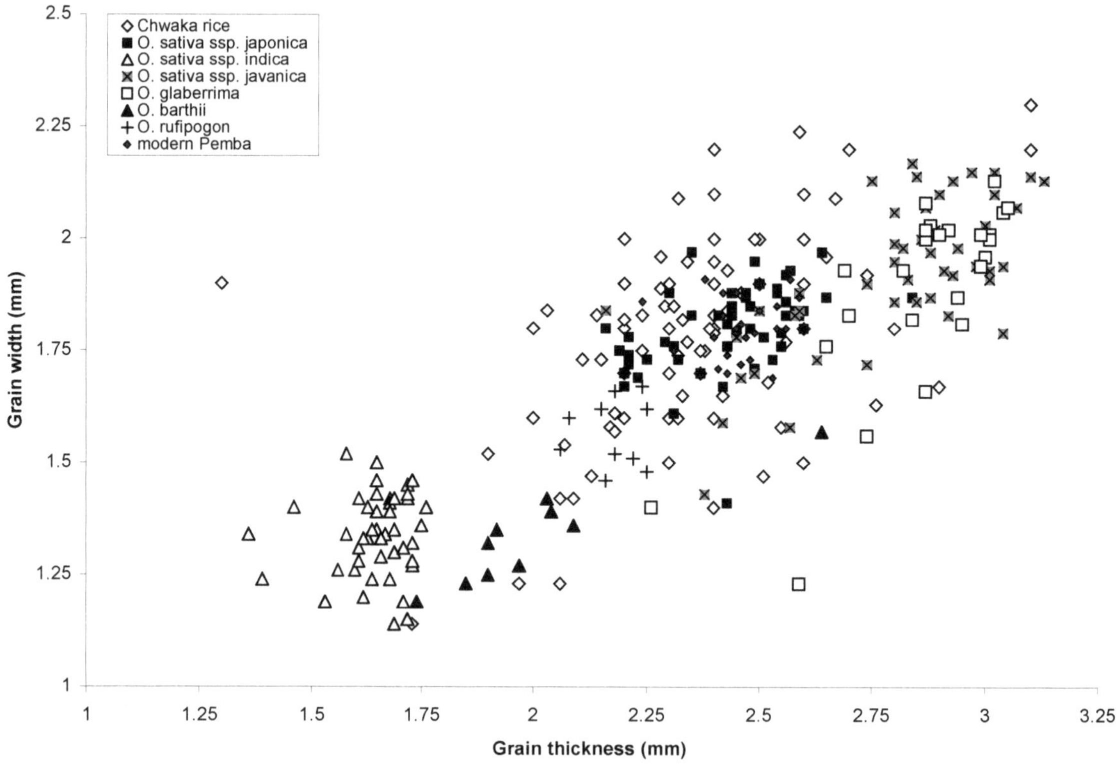

Figure 6.7. Archaeological (carbonized) rice from Chwaka compared with uncarbonized modern rice species including African domesticated rice (*O. glaberrima*) and its wild progenitor (*O. barthii*), and the Asian (*O. sativa*) varieties japonica, indica and javanica and a putative wild ancestor (*O. rufipogon*), using the metric variables of grain width and grain thickness. Modern Pemban rice is also shown.

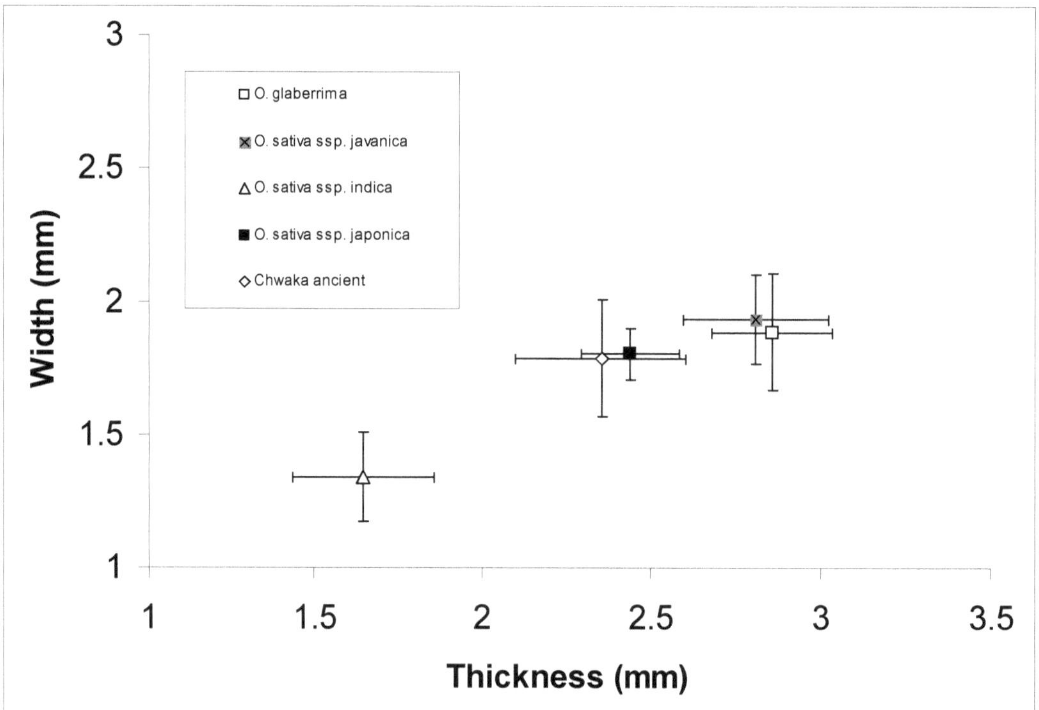

Figure 6.8. Mean breadth/width of carbonized Chwaka rice compared with means of uncarbonized modern rice varieties with standard deviation shown as error bars

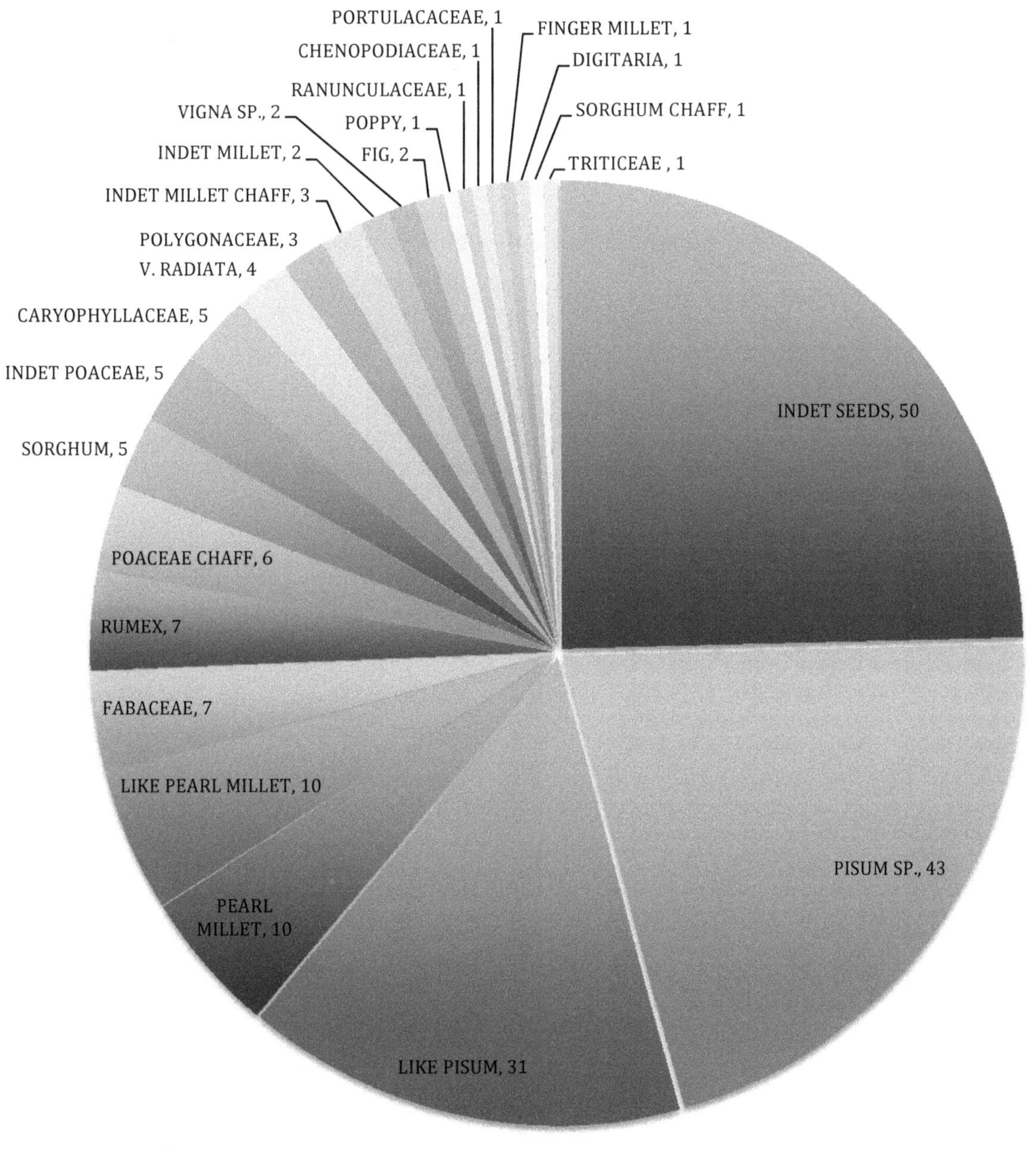

Figure 6.9. Frequency of seeds other than rice and cotton at Chwaka.

occupations of Chwaka during the fourteenth and fifteenth centuries A.D.). This indicates a broadening of the subsistence base to include African domesticates toward the end of Chwaka's occupation. One grain of finger millet was found at Chwaka (3% ubiquity), in a context that contained abundant rice but also one grain each of pearl millet and sorghum. Four probable but incomplete sorghum grains were distributed among three samples, and one glume fragment likely to be from sorghum appeared in another context, bringing the ubiquity for sorghum up to 11.8%. Two of the possible sorghum grains come from the mosque cistern, where the inhabitants of Chwaka may have dumped trash during the abandonment of the site. This is consistent with a return to millet agriculture toward the end of Chwaka's habitation.

Taken together, the evidence at Chwaka shows that African millets were known to Chwakans but that they did not form a staple of the economy. The fact that African millets appear more common towards the end of the occupation suggests that millets may have become popular once again, either due to drought years or because rice agriculture alone could not sustain the population.

Other Grains and Chaff. Two samples contain grass seeds that resemble members of the wheat tribe (Triticeae), although these are too fragmented to permit identification to genus. One *Setaria* grain was recovered, likely the Asian domesticate *S. italica,* although possibly the wild *S. megaphylla,* which grows wild in Ngezi forest (Koenders 1992:24). Although this millet was not found in any other Pemban site discussed here, a millet resembling *Setaria* was reported in Huffman's (in Wright 1984) archaeobotanical study on the Coromos islands, and thus is not unexpected in Swahili contexts. One *Digitaria* seed was found, probably a weedy grass species like *D. longiflora* or *D. milanjiani*, both present on Pemba (Koenders 1992:23). Two millet seeds too fragmented to permit identification were found at Chwaka, in addition to three fragments of chaff belonging to unknown millet species. Five unidentified members of the grass family were recovered, as well as six pieces of unidentified grass chaff from one context.

Straw fragments from unidentified grasses were found in 76% of Chwaka samples, averaging .01 g/liter in density. Although Op-2 provided some samples with large numbers of stems, they are widely distributed throughout the site, as are rice grains themselves. While Chwaka bears some stone structures, and the other sites do not, the discussion of stems as construction materials is still relevant because stone houses may have been roofed with thatching. Thus, the high frequency of stems at Chwaka may be related to roofing materials, crop production, local pottery production, or an unknown source.

Fabaceae. Beans, peas, or other members of the Fabaceae family were present in fourteen samples (41% ubiquity). The contexts that contained rice concentrations in Op-2B also contained numerous peas and probable peas, 72 in total (see Figure 6.10). The circumstances of deposition and preservation that resulted in the rice concentration probably led to the unusual presence of these peas as well. Legumes are relatively rare in charred archaeobotanical assemblages, possibly because they require little processing and are cooked in semi-liquid preparations, and thus do not get exposed to direct heat.

Four specimens of probable mung bean (*Vigna radiata*) of Indian origin were found, dispersed throughout the site (11.8% ubiquity). Two other bean specimens that could be assigned to the genus *Vigna* but not to species were also recovered. Chwaka samples also contained seven unidentifiable Fabaceae remains. Among these is a weedy legume, belonging to either the genus *Astragalus* or *Melilotus* (tribes Galegeae and Trifolieae respectively).

Cotton. Cotton (see Figure 6.11) was present in 30 samples (88% ubiquity), a major increase in frequency from Tumbe (30%) and Kimimba (absent). A total of 591 seeds were estimated, including an extraordinary cotton concentration found in Feature 192 (Op-2C) that contained 404 seeds estimated by weight (96.43 g).

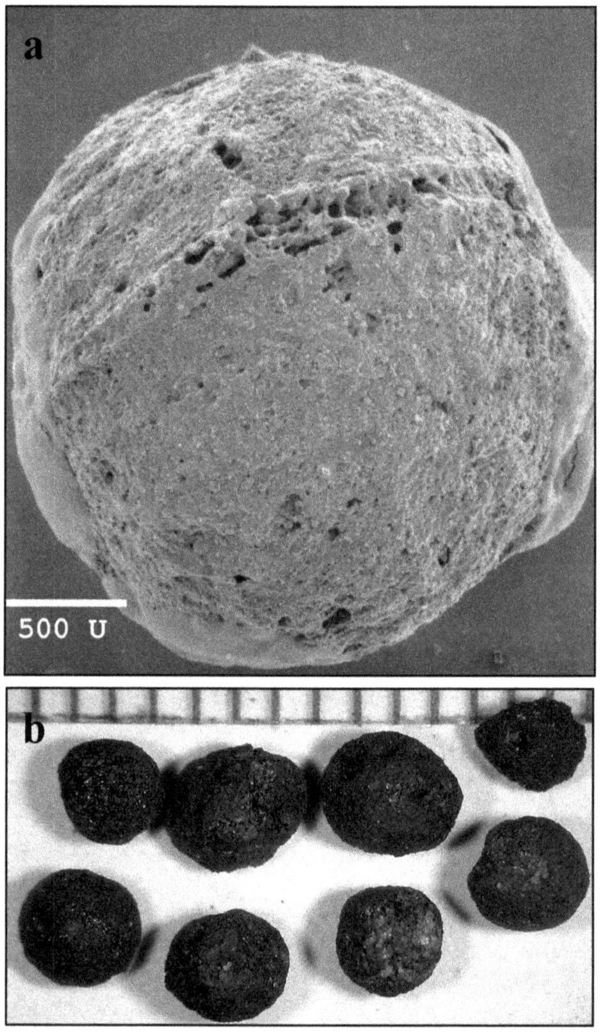

Figure 6.10. Archaeological peas from Chwaka: (a) scanning electron micrograph (scale in μm); (b) light microscope image (scale in mm).

This is a major increase in cotton use at Chwaka compared with earlier time periods, and can be reasonably interpreted as signaling an increased reliance upon cotton products in the local economy, although it could also be a result of the favorable preservation conditions existing at Chwaka overall. Here, dense middens accumulated due to the high population contributing to them, and deposits were probably buried quickly, securing excellent preservation conditions. The rice and cotton concentrations noted above were found within house features, and as such may have resulted from an accidental burning of stored materials in a fire. But considering that spindle whorls were common finds at

Chwaka, as they were for Shanga at the same time period (Horton 1996:396, Table 30), it is reasonable to interpret the cotton finds as indicative of intensified cotton production at Chwaka.

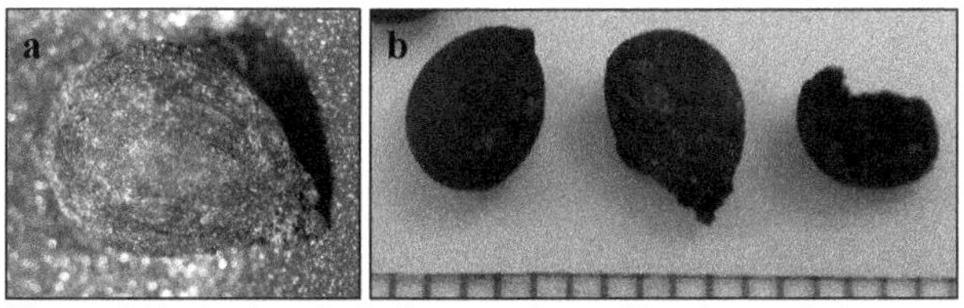

Figure 6.11. Archaeological cotton from Chwaka, scale in mm. Note linty residue toward cap on specimen in (a) and middle seed in (b).

One whole cotton seed (depicted in Figure 6.11a) was submitted for AMS radiocarbon dating, which returned a date (Beta-202826) of 890 ± 40 B.P., or calibrated A.D. 1030-1240. This falls within the early occupation of Chwaka, and provides a date for the cotton storage feature in Chwaka Op-2C (F-192) occurring at one of the lower levels (Layer 12).

The presence of abundant whole cotton seeds affords a rare opportunity to verify species. The observation of lint preserved on some of the specimens indicates that the cotton on Pemba was most likely domesticated, although other members of the cotton tribe Gossypieae also bear hairy seeds (Fryxell 1979). Of the three Gossypieae reported for Pemba (Koenders 1992:32-33), only domestic old world cotton (*Gossypium herbaceum* or *G. arboreum*) fits with the seed size and presence of hairs found in the archaeological specimens (Fryxell 1979; Vollesen 1987). Based on seed shape, the Chwaka specimens resemble *G. herbaceum* more so than *G. arboreum*. This is not unexpected; *G. herbaceum*'s wild progenitor (*G. herbaceum* var. *africanum*) has distribution today only in southern Africa, so it was likely domesticated in Africa. The earliest archaeological evidence for possible *G. herbaceum*, however, comes from Harrapan period sites in Pakistan, including by 2250-1750 B.C. at Mohenjo-Daro (Fuller 2000; Zohary and Hopf 2000). Conversely, *G. arboreum* is distributed further north and east within Asia, with cultivation in China dating to A.D. 900, although it was known to cultures in the Near East by 600 B.C. (Fryxell 1979:169-179; Zohary and Hopf 2000:135).

Other. The Chwaka mosque cistern sample contained one seed of the opium poppy, *Papaver somniferum*. This was the only find of this taxon across all sites reported. Opium poppy is a central Asian domesticate, and its presence at Pemba should not be surprising given the volume of trade connecting it with Asia. Poppy seeds could have been used as a spice or condiment, or the plant may have been grown for the opium, a pain-relieving drug contained in latex harvested from immature capsules (Lewis and Elvin-Lewis 1977:161).

Seeds– Wild & Unidentified
Only 26 wild or weedy seeds were recovered from among all the Chwaka samples, representing a minimum of seven taxa. This low density of weedy taxa is comparable to the values reported above for Tumbe and Kimimba. Of note is the distribution of weedy taxa: none of the samples from Op-1 or Op-2 contain weed seeds.

Two possible fig seeds were recovered, which suggests a continuation of the fig consumption activities indicated by fruit and seed portions found at Tumbe and Kimimba. The remaining wild/weedy taxa are common agricultural weeds: dock (*Rumex*, n=7), unidentified members of the dock family (Polygonaceae, n=3), buttercup family (Ranunculaceae, n=1), pink family (Caryophyllaceae, n=5), purselane family (Portulacaceae, n=1), and goosefoot family (Chenopodiaceae, n=1). Their presence in low abundance at Chwaka is not surprising, considering that agricultural weeding activities take place in fields, which at Chwaka were in all likelihood close to, but not actually within, the town itself. Furthermore, the harvesting of rice by hand greatly reduces the inclusion of weeds, as I observed in recent ethnobotanical studies on Pemba. Thus weeds would not be an expected by-product of agricultural activities, particularly in urban contexts.

Unidentified seeds. Fifty-eight percent of samples contained a total of 50 unidentified seeds, many of which were likely fragmentary or eroded domesticates lacking the portion allowing a positive identification (i.e., beans, cotton endosperm).

Non-Botanical Categories
Some of the diverse artifacts known from Chwaka appear in archaeobotanical samples, typically in the heavy fractions, but often in the light fractions as well. Non-fish faunal remains were recovered from 16 samples (47.1%) at an average density of .03 g/liter. With a few exceptions, Operations 1 and 2 were low in animal bone. Shell was identified in only three contexts, including a significant amount from the mosque cistern sample. Fish bone, sometimes in large quantities, was found in 67.6 % of samples – these were set aside for future analysis by a fish specialist. Most fish elements were scales and vertebral/spinal bones, but some cranial and gill elements were also recovered, with a mean density of .15 g/liter. Coral was discovered in fifteen samples, which is not surprising because coral was used in constructing stone buildings (see figure 6.12), and forms a sedimentary geological stratum close to the surface. Undiagnostic pottery fragments were present in 61.8 % of samples and diagnostic sherds were present in three (8.8%) samples. Taking the pottery and daub data together, all ceramics were present at an average density of .11 g/liter.

Figure 6.12. Coral architecture on Pemba: (a) Haruni's tomb, a standing monument near the Chwaka mosque, (b) a closeup of non-porites coral used in wall construction.

Twenty-seven samples (79.4%) contained residual sediment concretions or quartz pebbles. One small piece of blue glass was noted in Bag 265 (Unit 11) along with several possible pieces of iron slag. One small glass bead was discovered in Bag 1453 (Op-1A). Flotation samples, particularly heavy fractions, were the source of several beads that were returned to the artifact lab during flotation, and would likely have been too small to be recovered using ¼-inch window screen commonly used to sift dirt at the Pemban excavations.

KALIWA– Fourteenth to Fifteenth centuries A.D. (n=11)

Results

Kaliwa was an earthen village site excavated in 2004 by students participating in a field school sponsored by the University of Dar es Salaam's Archaeology Unit and by the Pemba Archaeological Project. Five test units were excavated, and four were found to contain significant artifact concentrations, particularly Unit 2 (a house floor) and Unit 5 (a dense midden). While only 11 light fraction and one heavy fraction samples from Kaliwa have been analyzed to date, these represent all units bearing cultural material and much of the time span of Kaliwa, and thus serve as a good representation of Kaliwa in the Pemban archaeobotanical record. The mean flot density at Kaliwa was 3.93 g/liter. Unlike the other sites, where flot density was skewed by a few wood-heavy or rice-intensive samples, Kaliwa samples were more even, ranging from 1.82 g/liter to 7.40 g/liter. Unit five bore the heaviest samples overall. This density is on par with samples from Chwaka, with the exception of the uncommonly rich Op 2. Density of identifiable plant remains averaged 1.46 g/liter, which is somewhat lower than Chwaka units other than Op 2. As will be discussed below, the high identifiable flot density found with Unit 5 derives from the high abundance of fruit and nut fragments found there. Archaeobotanical frequencies of seeds and weights and densities of non-seed categories are presented in Table 6.3.

Table 6.3. Kaliwa Archaeobotanical Results

	Unit 2	Unit 3	Unit 4	Unit 5		Site Totals
Samples (n=)	3	2	2	4		
Volume floated (l)	30	22	20	39		
					Density	Ubiquity
Non-Seed Items					(g/l)	(%)
Wood (g)	5.16	3.72	1.48	6.90	.16	100
Parenchyma (g)	.36	.27	.06	.13	<.01	73
Stem (#)	9	4	3	28	<.01	91
Coconut (g)	.38	.16	.10	1.19	.02	100
Fruit/Nut (g)	6.13	1.98	2.52	22.65	.34	100
UNID (g)	1.00	.64	.15	.31	.02	82
Uncarbonized (g)	15.99	18.60	3.10	12.14	.43	100
					Seed Total	Ubiquity
Seeds & Chaff - Domesticates						(%)
Pearl millet				4	4	18
Finger millet	1				1	9
Rice	3	1	9	8	21	82
Rice chaff	8	7	46	10	71	91
UNID millet				1	1	9
UNID millet chaff		1			1	9
UNID Poaceae				1	1	9
Cotton	5	4	4	17	31	100
Seeds & Chaff – Wild						
Mallow family	1				1	9
Legume family				1	1	9
Dock family		1		1	2	18
Pink family				3	3	18
Purselane family		1			1	9
Unit Seed Total	*10*	*8*	*13*	*36*	*67*	*100*

Preservation and Disturbance
Uncarbonized plant fibers (largely roots) and seeds were found in all samples, with a mean density of .43 g/liter. This pattern is more similar to that found at Tumbe than either Chwaka (the contemporaneous site) or Kimimba (an earlier village site). Uncarbonized seeds were not counted by taxon; only a total count per sample was noted, therefore I do not present frequencies of uncarbonized seeds here. Uncarbonized seeds included dock (*Rumex*), purselane (*Portulaca*), mustard (*Brassica* sp.), crowfoot grass (*Dactylotenium)*, amaranth, and members of the grass, legume, aster, tomato, sedge, mallow, and pink families. These are common weeds of cultivation, which is interesting considering that today Kaliwa lies on an uncultivated path toward the fish market just south of the modern town of Tumbe. The uncarbonized weed assemblage may derive from the disturbed environment adjacent to this path, or it may signal cultivation of crop plants or encouragement of wild plant stands here in the recent past.

Wood
Wood was present in all samples in relatively even distribution across the site (see Table 6.3 and Figure 6.13 for Kaliwa ubiquity data). The average wood density at Kaliwa came to .16 g/liter. This figure is lower than that recorded for Chwaka, but comparable to wood densities at Tumbe and Kimimba. This was the only Pemban site for which wood was not the dominant taxon by weight. Wood at Kaliwa comprised the third most abundant category as measured by density relative to sample volume (see Figure 6.14 for Kaliwa density chart). As described previously, no attempt was made to identify any of the Kaliwa wood fragments to taxon.

Storage Tissues
Seventy-three percent of samples contained parenchyma, but in low abundance (<.01 g/liter mean density). This is similar to the ubiquity of parenchyma at Kimimba and Chwaka, but more than documented for Tumbe. As reported above and in the previous results chapter (Chapter 5), parenchymal tissues may derive from the underground storage organs of common crops (e.g., yam, taro), or from fruits (e.g., banana).
None of the samples contained fragments attributed to roots/tubers, which is consistent with their infrequency at Chwaka. In this regard, Kaliwa and Chwaka are more alike than either one is to the earlier sites reported in Chapter 5: both Tumbe and Kimimba produced well preserved root/tubers in some samples.

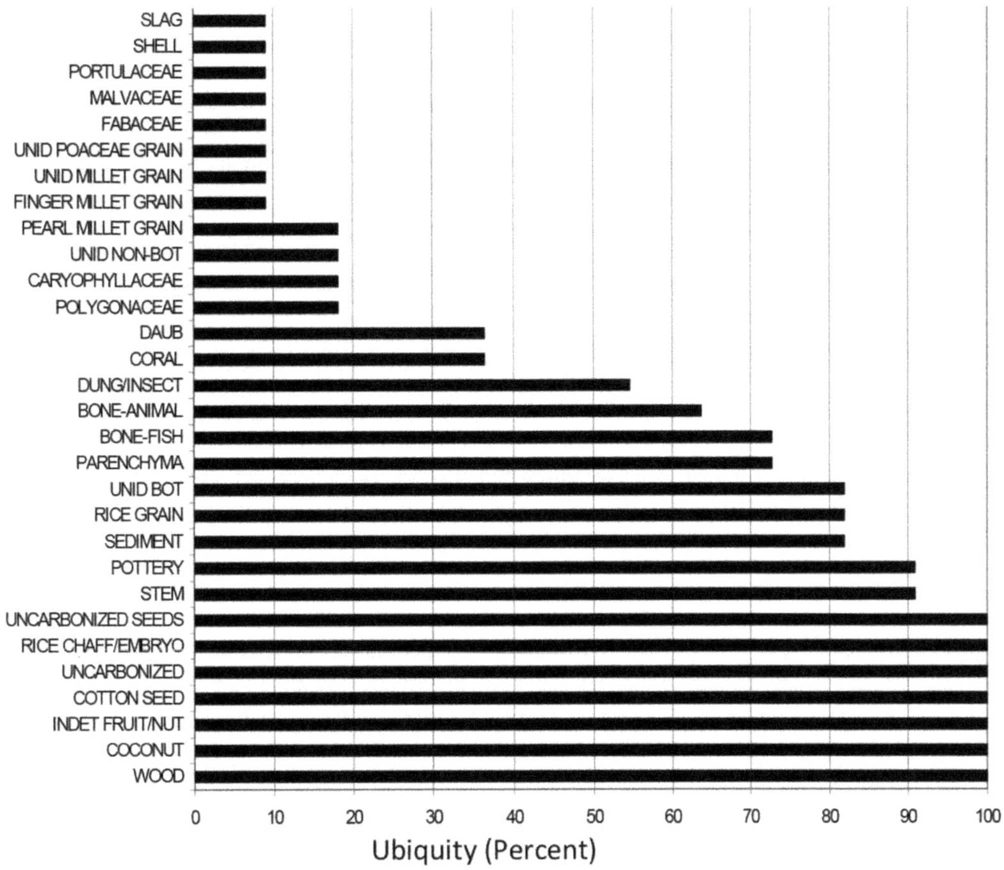

Figure 6.13. Ubiquities for Kaliwa sorted materials ≥2 mm plus identified seeds.

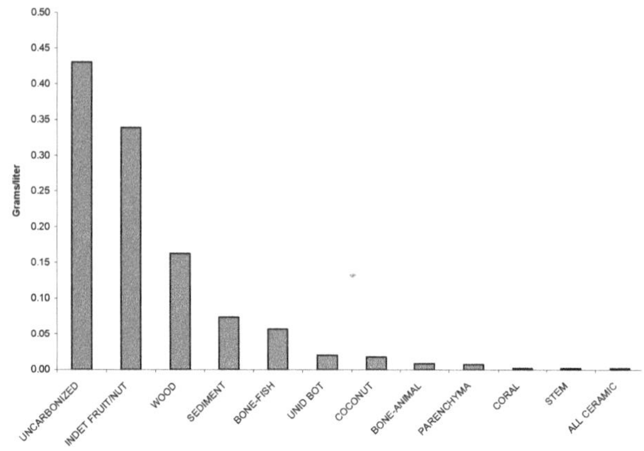

Figure 6.14. Densities of materials ≥2 mm at Kaliwa by weight (g/liter).

Coconut and Relatives

All Kaliwa samples contained coconut fragments, similar to Chwaka (91%), where coconut was more ubiquitous than at either Tumbe (78%) or Kimimba (41%). The mean density of coconut at Kaliwa was found to be .02 g/liter, which is on par with coconut densities at Tumbe and Kimimba but lower than Chwaka's coconut density by a factor of five. Other specimens resembling the coconut tribe, possibly the oil palm, were found in 4 (36.4%) samples for a total of 29 fragments. This frequency of non-coconut palm is similar to that reported above for Chwaka, but much higher than at Kimimba (6%) and Tumbe (absent).

Other Fruit/Nut

Nut fragments were relatively abundant in Kaliwa samples, comprising the second most abundant category behind uncarbonized remains. Fruit and nut pieces were found in all samples, as at Chwaka, which is slightly higher than the values reported for the first millennium sites. At an average density of .34 g/liter, fruit and nut fragments were as abundant at Kaliwa as at Chwaka, but more than Tumbe and Kimimba. Two distinct yet unidentified nut types were found at Kaliwa that were absent from the other sites (see Figures 6.15, 6.16), and further study of the Kaliwa fruits and nuts are expected to contribute even more nut types to the Pemban archaeobotanical record.

UNID Botanical Material

Most samples (81%) contained a significant number of unidentifiable botanical specimens, and in one sample unidentified botanical remains were more abundant than any other category. The average density of unidentified plant pieces was calculated to be .02 g/liter, making this category the sixth most abundant by density. The frequency of unidentified botanicals at Kaliwa is consistent with that reported for the other sites, although the mean density is lower than that of Chwaka.

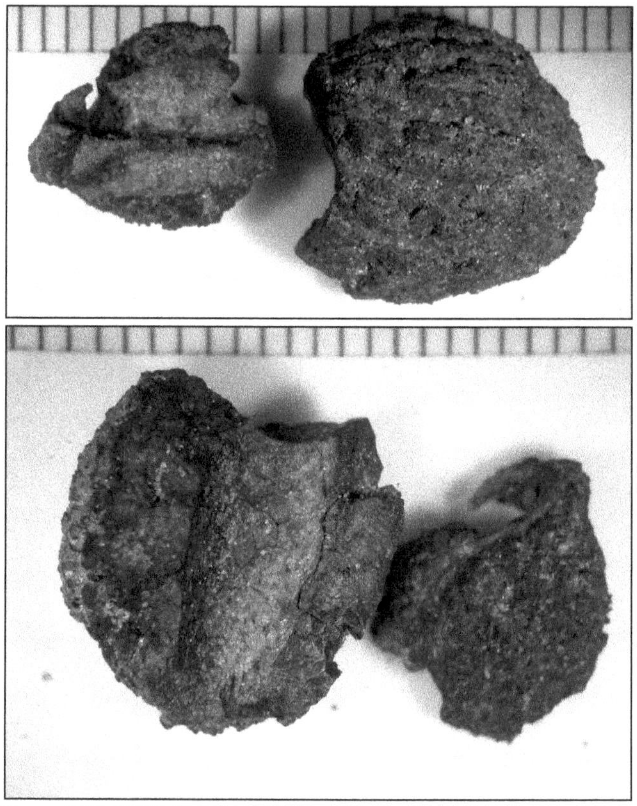

Figure 6.15. UNID fruit/nut "C" from Kaliwa. Scale is in mm.

Figure 6.16. UNID fruit/nut "D" from Kaliwa. Scale is in mm.

Seeds and Chaff – Domesticates

African millets. A total of five grains of millet were found at Kaliwa, not including one grain and one glume from unidentified millets (see Figure 6.17 for seed representation by abundance). Four grains of pearl millet were recovered, three from one context in Unit 5 (most recent layer), which interestingly was the only context from the entire site to lack rice rachillae. One finger millet grain was found, also from the upper level (see Figure 6.18). This low frequency and abundance of African grains accords with

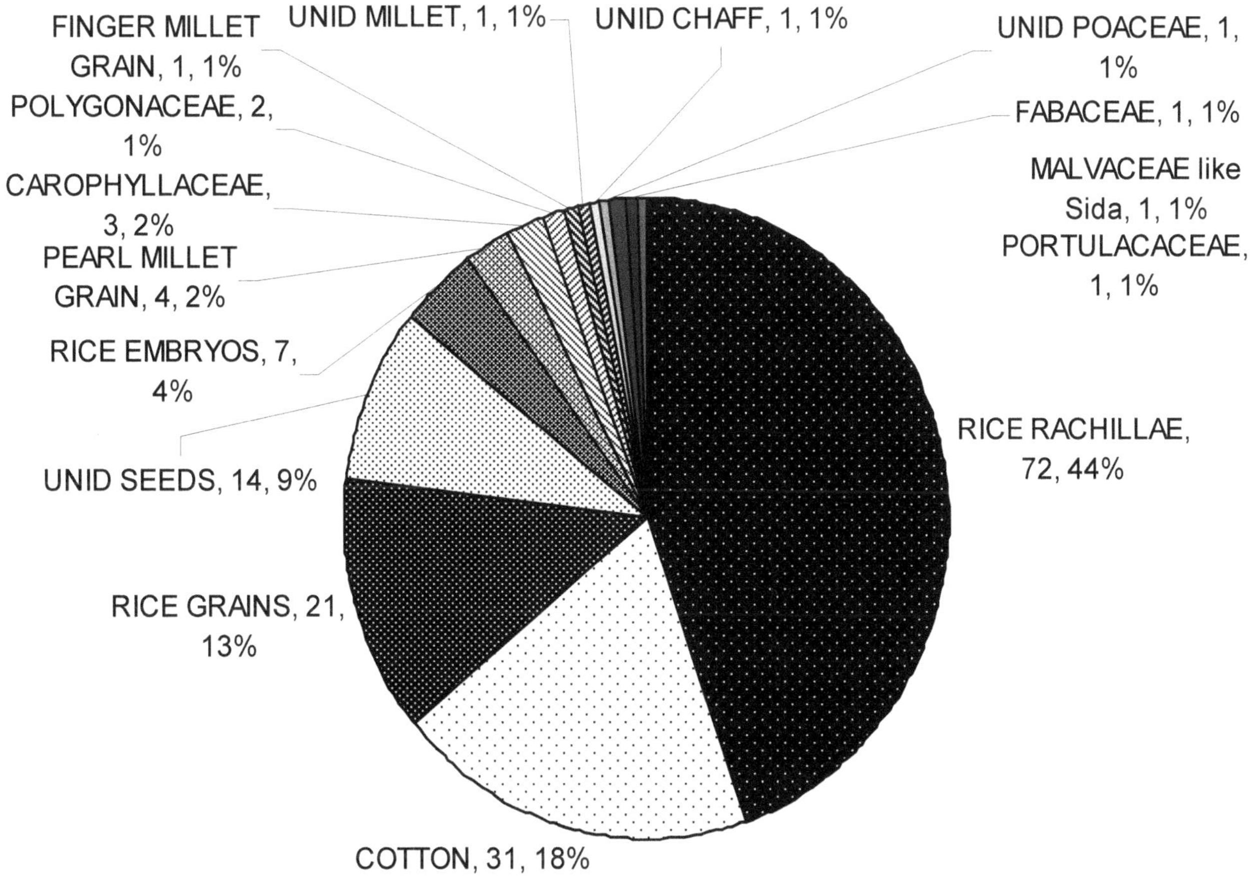

Figure 6.17. Seed and chaff frequencies at Kaliwa.

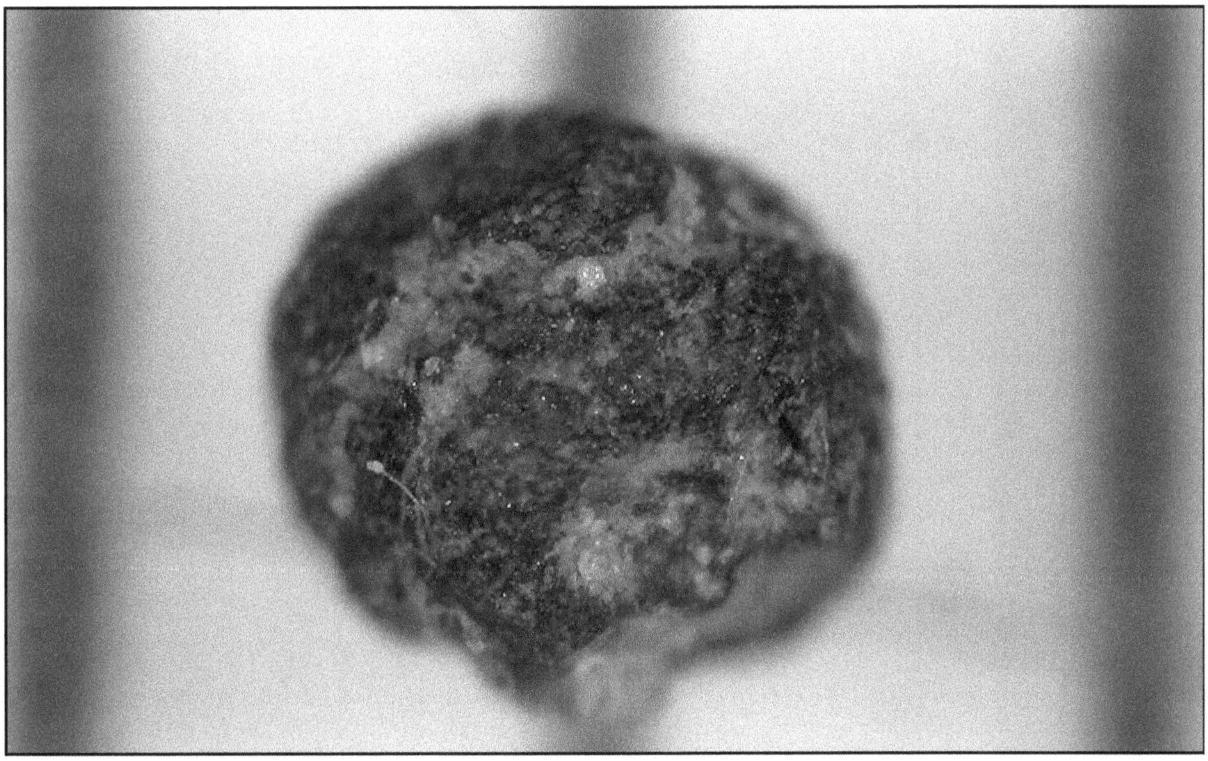

Figure 6.18. Finger millet grain from Kaliwa. Scale in mm.

the pattern documented above for Chwaka, as does the skewing of frequency toward the most recent time period, probably during the 15th century.

Rice. All samples from Kaliwa contained some element of the rice plant such as seeds, embryos, or rachillae. Nine (82%) samples contained rice grains, mostly numbering under three grains except for one richer sample, which contained seven grains. Ten (91%) samples contained rachillae of rice, with two samples from Unit 4 containing 64% of the rachillae for the entire site. Freed rice embryos were located in four samples (36.4% ubiquity). Rice grains were relatively rarer than rachillae, in both frequency and abundance measures, following the trend observed among other Pemban archaeobotanical samples.

Cotton. Fragments of cotton seeds were found in all of the Kaliwa samples (100% ubiquity), but in relatively low abundance (31 seeds total; average 2.8 seeds/sample). This frequency is similar to but even higher than that reported above for Chwaka, and significantly higher than that found at either Tumbe or Kimimba.

Seeds –Wild and Unidentified

Only four samples contained weed seeds, totaling nine seeds identifiable only to the following plant families: grass (1); legume (1); mallow (1); dock (2); pink (3); and carpetweed (1). This low abundance and frequency of weed seeds is similar to that found at all other Pemban sites, a pattern which has been discussed above as reflecting the common agricultural practices. Six samples (54.5%) produced a total of 14 unidentified seeds. All samples contained unidentifiable seed fragments.

Non-Botanical Materials

All samples contained some variety of non-botanical artifacts that were sorted and counted if found in or above the 2.00 mm screen. Animal bone other than fish was present in 63.6% of samples at a mean density below .01 g/liter. Fish bone was slightly more frequent, occurring in 72.7% of samples, but much more dense than animal bone at .06 g/liter. In particular two samples from Unit 5 (layer 2) were especially heavy in fish remains, averaging .3 g/liter, and raising the mean density significantly for this category above that of Chwaka. This density of fish remains is consistent with Kaliwa's close proximity to the shoreline; indeed, this locale is close to an active docking area and marketplace for fishermen today. Coral was found in four samples (36.4%) averaging a density of less than .01 g/liter. This is indicative of the quantities of coral bedrock that outcrop at localities throughout the island. Only one piece of shell was recovered from the whole site, which is noteworthy considering the site's proximity to the waterfront, but parallels the low frequency of shell among the majority of Chwaka samples.

Pottery was recovered from almost all samples (90.9%), including two diagnostic sherds. Fragments of daub were found in only three (36.4%) samples, and only where coral fragments were absent, similar to the pattern described above for Chwaka. Combining pottery and daub into one ceramics density gives a measure of .02 g/liter, far below that of Chwaka. Small fragments of sediment or stones were found in most (81.8%) contexts, at a density of .07 g/liter. Several slag fragments were reported from one Unit 2 context. No beads or glass shards were recovered from the Kaliwa flotation samples.

TUMBE – Mazrui Settlement Eighteenth to Nineteenth Centuries (n=10)

The upper layers (Layer 1, and occasionally 2) of the site of Tumbe bear evidence of an occupation by a Mazrui Arab population in the eighteenth to nineteenth centuries. The Mazrui were a prominent Omani group that came to the coast in service to Omani rulers early in the 18th century and served as highly effective governors, particularly in Mombasa (Maxon 1994:50). The ten archaeobotanical samples from this stratum of Tumbe are presented alongside the other second millennium A.D. occupations, even though no temporal overlap is indicated, to provide a later contrast to the Chwaka and Kaliwa Swahili remains. Moreover, it is important to document the content of these layers that overlie the earlier Tumbe assemblage to determine what materials may have been introduced during possible intermixing.

Tumbe Mazrui flotation samples are dense in archaeobotanical remains (mean flot density is 5.19 g/liter), but much of this comprises woody tissue and uncarbonized plant remains. The identifiable flot density, here understood as the density of materials sorted at or above 2 mm, comes to 1.91 g/liter, 1.50 g/liter of which is represented by wood and uncarbonized materials. Figure 6.19 shows the relative abundance of items in the seed and chaff category.

Preservation and Disturbance

These flotation samples come from layers located close to the surface and therefore are more susceptible to surface contaminants than the lower Tumbe deposits. This is seen in the density of uncarbonized plant remains (.70 g/liter) and the ubiquity of uncarbonized remains (100%) and seeds (90%). Insect remains, insect eggs, dung, and fungus were also common components of these upper Tumbe layers, indicating post-depositional disturbance.

Wood

As with all other samples from Pemba, carbonized wood is present in all samples (see Table 6.4, Figure 6.20), and is the most abundant taxon by weight (.81 g/liter; see Figure 6.21). One sample was particularly rich in wood (3.91 g/liter), particularly woody tissues that were layered and tubular, resembling small branch attachments. As discussed previously in this chapter and Chapter 5, the most parsimonious explanation for the abundance of wood in or near domestic contexts is the disposal of wood used as fuel.

Table 6.4. Archaeobotanical Frequencies from Mazrui Period Tumbe.

	U.13	U.16	U.44	U.53	U.60	U.63	U.64	OP-4E	Site Totals	
Samples (n=)	1	1	1	1	1	1	3	1	10	
Volume (liters)	18	16	10	12	4	1	29	10	100	
Non-Seed									Density (g/liter)	Ubiquity (%)
Wood (g)	4.25	2.93	.15	3.13	.59	2.06	37.31	1.54	.81	100
Parenchyma (g)	.07			.01		.08	.1	.16	.01	60
Stem (#)	6			5	7		37	1	<.01	60
Root/tuber (#)		3	1		7		4		<.01	40
Coconut (g)	.38	.83	.08	.14	.01		1.15	.13	.03	90
Fruit/nut (g)	1.33	.05	.29	.60		.01	.81	.24	.03	90
UNID (g)	.55	.21	.07	.13	.01	.01	.80	.01	.02	100
Uncarbonized (g)	6.21	10.97	2.92	7.74	2.43	.21	25.72	15.99	.70	100
Seeds & Chaff									Total	Ubiquity (%)
Pearl millet	2	1					2		5	40
Indet millet					3		1	1	5	30
Rice		4			1			2	7	30
Rice chaff		11		1	1		2	1	16	60
UNID Poaceae							1		1	10
Cotton		1	2				1	1	5	40
Legume family								1	1	10
Dock				2			2		4	20
Dock family		1							1	10
Spiderwort family				1					1	10
Carpetweed							3		3	20
Purselane family		1							1	10
UNID Seed A	1						1	1	3	30
UNID Seed F				1					1	10
UNID Seeds	7	6		2	4	1	20	2	42	80
*Unit Seed Total**	*10*	*14*	*2*	*6*	*8*	*1*	*31*	*8*		

* Excludes chaff frequencies

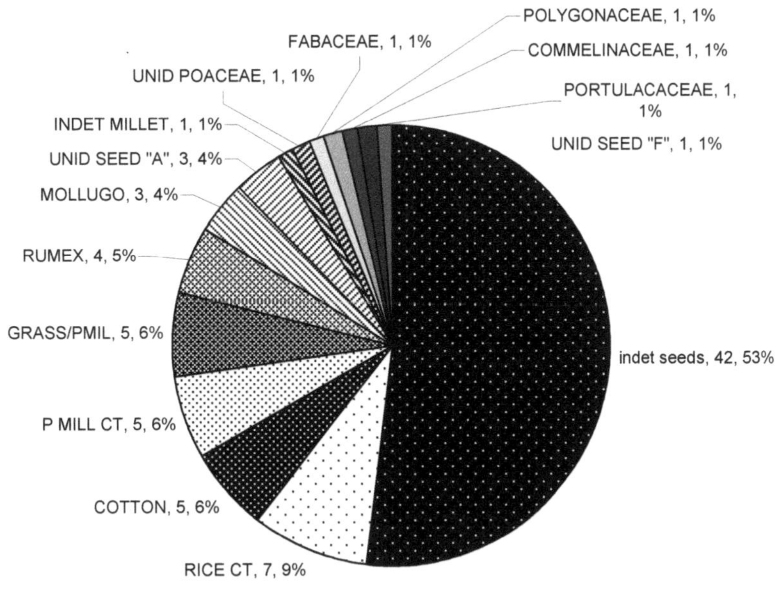

Figure 6.19 Mazrui Tumbe Seed Frequencies.

Storage Tissues
Parenchymal tissues were found in 60% of the Tumbe Mazrui samples, comprising a density of .01 g/liter. This is a higher frequency of parenchyma than the lower Tumbe samples, and a slightly higher density than the other sites discussed above. Roots or tubers were present in 40% of the samples from Mazrui Tumbe, including three small underground storage organs resembling those found in the sedge family The presence of root and tuber tissues is more similar to Tumbe and Kimimba, as opposed to Chwaka and Kaliwa, where they are absent.

Coconut and Relatives
Coconut endocarp is found in 80% of samples, weighing an average of .03 g/liter of sediment floated. A fragment of possible exocarp (husk) was recovered from an additional sample, bringing the overall coconut

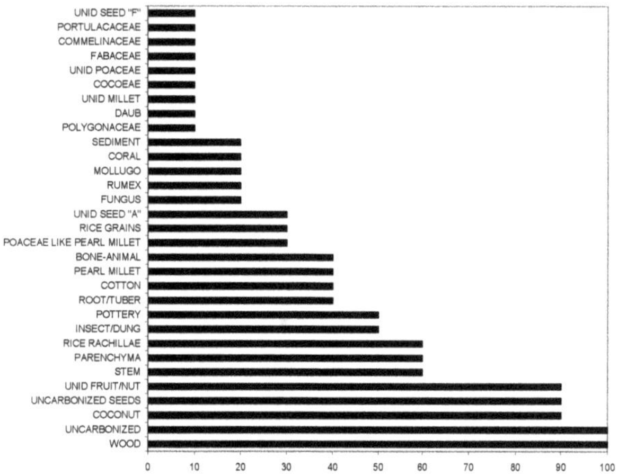

Figure 6.20. Mazrui Tumbe ubiquity measures.

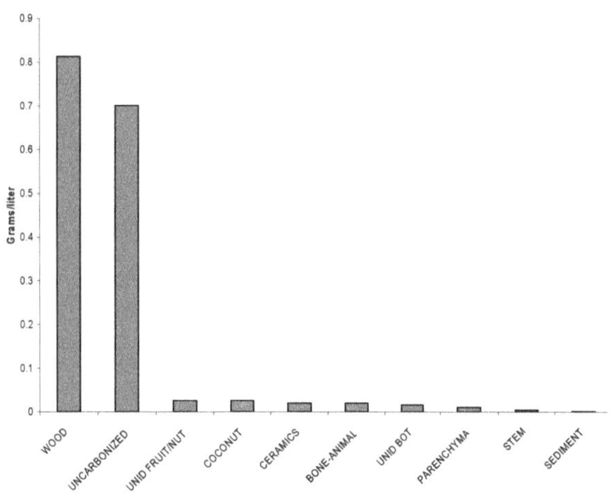

Figure 6.21 Density of Mazrui Tumbe Materials ≥ 2 mm.

ubiquity measure to 90%. The frequency of coconut is similar to all the sites except Kimimba, and the abundance as measured by density is similar to all sites except Chwaka, where coconut was much denser. In addition, two fragments of a member of the coconut tribe were recovered.

Other Fruit/Nut
Unidentified remains of non-palm fruits and nuts were very common at Tumbe during the Mazrui period. They were present in 90% of samples, and weighed on average .03 g/liter, the same as coconut. This lower density is more in keeping with the results discussed for the first millennium sites in Chapter 5, because this category was more abundant in samples from Kaliwa and Chwaka.

UNID Botanical Material
All samples contained some carbonized plant material that was unable to be assigned to one of the categories above. Unidentified plant remains were present, on average, at a density of .02 g/liter, which is much more than the lower Tumbe strata, but in keeping with values from Kimimba, Kaliwa, and Chwaka. were present in 40% of the samples from Mazrui Tumbe, including three small underground storage organs resembling those found in the sedge family. The presence of root and tuber tissues is more similar to Tumbe and Kimimba, as opposed to Chwaka and Kaliwa, where they are absent.

Seeds and Chaff – Domesticates
Pearl Millet. The only African grain present at Mazrui Tumbe was pearl millet, which was present in 4 samples (40% ubiquity) for a total of 5 grains, not including 5 grains of possible pearl millet found in 3 samples (see Figure 6.19 for graph of relative seed abundance). No millet chaff was recovered. The frequency of pearl millet is similar to that found in the other 2nd millennium sites discussed in this chapter, in contrast with underlying Tumbe levels (74%) and Kimimba (5.9%).

Rice. Fragments of rice were recovered from three (30%) samples, totaling seven estimated rice grains. Following the dominant pattern observed among the previous Pemban sites reported here, rice rachillae are more abundant (16 total) and more frequent (60%) than rice grains. Interestingly, the context with the most rice grains (U.16) also had the most rachillae, in a ratio of 1:2.8.

UNID Poaceae. One unidentified millet was found in the Tumbe Mazrui samples, resembling *Setaria*, an Asian millet. One small, probably weedy, grass seed was also recovered. Stem fragments categorized as probable chaff or thatching were present in 60% of samples, at a density below .01 g/liter.

Legumes. Only one Fabaceae seed was located among the Tumbe Mazrui samples, and this could not be identified to genus.

Cotton. Cotton is much rarer at Mazrui Tumbe compared with the high frequency of cotton seen at Chwaka and Kaliwa. Fragments of cotton seeds were present in four (40%) samples for a total of five seeds across the site, similar to the cotton pattern observed at Kimimba and in lower Tumbe levels. This low incidence of cotton seeds may be due to an increased availability of manufactured textiles through local and long-distance trade, or it may reflect overall poorer preservation conditions existing at Tumbe.

Seeds – Wild

Six (60%) samples contained wild or weedy seed types; 14 wild seeds in total were recovered. These included dock (4), *Mollugo* (carpetweed – 3), UNID seed "A" in the poppy family (3), UNID seed "F" (see figure 6.22, resembling *Convolvulus* – 1), and members of the dock (1), spiderwort (see Figure 6.23 - 1), and Portulacaceae (1) families.

Figure 6.22. UNID seed "F", likely from the morning glory (Convolvulaceae) family. Scale is in mm.

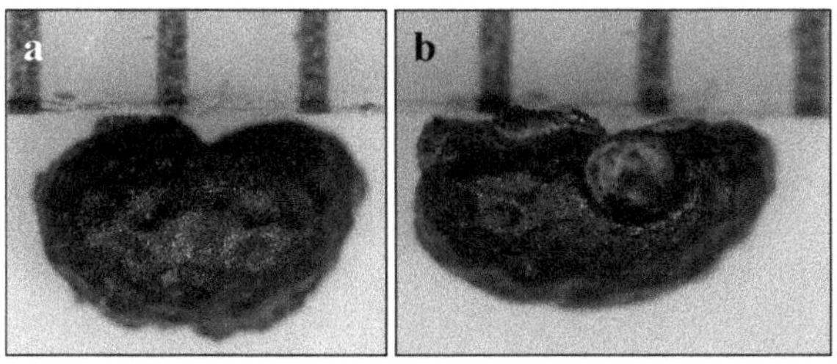

Figure 6.23. Seed in the spiderwort (Commelinaceae) family; (a) dorsal view, (b) ventral view. Scale is in mm.

This relatively low incidence of weed seeds places Mazrui Tumbe levels alongside the other sites mentioned above is best interpreted as signaling hand-harvesting practices, which eliminate weeds from the harvest. Nine (90%) samples produced unidentified seeds, numbering 42 in total. The same frequency is observed for unidentifiable seed fragments.

Non-botanical

Mazrui Tumbe samples were not generally rich in non-botanical artifact types greater than or equal to 2 mm. Animal bone was present in 40% of the flotation samples, averaging .02 g/liter, but no fish remains were recovered. This pattern is quite different from the fauna of other second millennium archaeobotanical samples, where fish comprises the most abundant faunal find. This aspect of Mazrui Tumbe's archaeobotanical assemblage most resembles the lower Tumbe layers, indicating perhaps a preservation bias in action here. Coral fragments were found in two (20%) samples, and no shell was located.

Sherds of pottery were found in 50% of samples, and none were diagnostic. Daub was present in only one sample (from Unit 64), where it was found in abundance (26 fragments; .42 g). Taken together, all the ceramics at Mazrui Tumbe were represented at a density of .02 g/liter. Sediment chunks were present in only two samples.

Summary – Second Millennium Archaeobotanical Patterns

This chapter contains a detailed archaeobotanical record for the sites of Chwaka and Kaliwa for the purpose of comparing patterns between this Swahili stonetown and village. While Chwaka only overlaps with Kaliwa toward the end of its occupation, the comparison between the two at this important juncture, in the fourteenth and fifteenth centuries, remains informative. The Tumbe data from the Mazrui period is not directly related to the earlier food production system taking place at Chwaka, but plant remains during this latest occupation are instructive both for understanding late pre-colonial foodways and for determining potential sources of contamination for the underlying seventh to tenth century Tumbe deposits.

Non-Seed Components

While wood is ubiquitous at both Chwaka and Kaliwa, its density at Chwaka is threefold of that at Kaliwa, where it is only the third most abundant material. This disparity in wood dominance, coupled with the relatively high abundance of other fuel sources such as coconut endocarp and other fruit/nut fragments, may indicate that Chwaka was able to command a greater share of the wood available by the fourteenth century. With respect to preserved plant storage tissues such as parenchyma and roots/tubers, Chwaka and Kaliwa display similar pattern: many samples bear small amounts of parenchyma, and only a few root/tuber fragments were retrieved from Chwaka, while none were present among Kaliwa samples. Coconut was almost omnipresent in samples from both sites, although it was more abundant at Chwaka by almost five times. Unidentified palm fragments were also common at Chwaka and Kaliwa, unlike Tumbe and Kimimba, where they were absent and rare, respectively.

Chwaka and Kaliwa were most similar in the class of unidentified fruit and nut remains. Both sites demonstrated 100% ubiquity for this category, and unlike any other material, Kaliwa surpassed Chwaka in the density of UNID fruit/nut remains, albeit narrowly (.34 g/liter compared with .32 g/liter). Moreover, Kaliwa showed evidence of a greater variety of taxa in this category and provided two novel fruit types ("C" and "D", shown above).

Seeds and Chaff

The seed records of Chwaka and Kaliwa are impressively alike, particularly with respect to taxonomic presence and frequency. Rice dominated the seed and chaff fractions of both Chwaka and Kaliwa; similarly, African grains were present in low abundance at each site. Clearly, the enormous quantities of rice grains and rachillae coming out of several Chwaka Operation 2 samples were unique to this house structure at Chwaka, and are likely due to accidental burning of food stores. The later levels of Chwaka, however, accord well with the patterns observed at contemporaneous Kaliwa. Generally, rice rachillae were more frequent and abundant than grains or other chaff components at both Chwaka and Kaliwa. Thus while based on our excavation sample, Kaliwa residents may not have stored rice in the quantities available to the residents of Operation 2 at Chwaka, their domestic use patterns are very similar to those found elsewhere at Chwaka, particularly those from the same time period.

All three African millets found on Pemba were recovered from Chwaka, while all but sorghum were recovered from Kaliwa. These were all found in relatively low frequency and abundance at both Chwaka and Kaliwa. Of particular interest is the distribution of African grains largely in the later sequences at Chwaka, those likely overlapping in time with Kaliwa. This return to African crops may signal a decrease in the success of or access to rice yields, including the possibility that rice was shipped elsewhere as a trade good, tax, or tribute.

Cotton shows a pattern similar to that observed for rice. Cotton is virtually 100% ubiquitous at both sites, occurring even more frequently at Kaliwa than at Chwaka. Cotton seeds occur in relatively small quantities except for one storage context found at the base of Operation 2 at Chwaka. This is significantly more cotton than was found at either of the two first millennium sites, and probably signals continued local production of cotton fibers for textile production destined for local use. Spindle whorls fashioned out of stone or pottery were found at both Chwaka (common) and Kaliwa (n=4), whereas they were notably absent from Tumbe and Kimimba, in keeping with other sites along the coast. The presence of spindle whorls, coupled with the increased frequency of cotton finds, can be reasonably interpreted as the introduction of or an increase in the local production of cotton textiles for clothing.

One striking difference between Chwaka and Kaliwa rests in the relative abundance of members of the pea/bean family. Weedy Fabaceae and several cultivated bean and pea taxa were significant contributors to the non-rice seed assemblage at Chwaka. Moreover, peas constituted a small but well preserved fraction of the samples responsible for the rice concentrations found at Chwaka's Operation 2. Domesticated beans and peas are entirely absent from the Kaliwa archaeobotanical record, and only one weedy Fabaceae was found. As discussed previously, legumes tend not to preserve well archaeologically, and thus their absence at Kaliwa is only notable because of their relative abundance at Chwaka, which may be owing to better preservation conditions existing at Chwaka, and to the extraordinary burnt storage feature. Weed seeds were quite rare at both Chwaka and Kaliwa, a pattern now familiar after observing it across all the archaeological sites on Pemba.

Mazrui Tumbe in Spatial and Temporal Context

One of the most interesting aspects of the Mazrui archaeobotanical record from Tumbe is how similar it is to the other Pemban sites, despite its cultural and temporal removal from the earlier Swahili settlements. While European and North and South American plants likely enjoyed a wider distribution in eastern Africa during Mazrui times, they do not appear in the archaeobotanical assemblage from the upper Tumbe levels. In particular, maize is notably absent, despite its probable introduction by this time, as are tomatoes, peppers, and other members of the nightshade family.

Like the other Pemban sites, Mazrui Tumbe samples were generally rich in wood, coconut, and unidentified fruit/nut remains. Both pearl millet and rice are represented, although rice dominates in frequency due to the wide distribution of rachillae. Stem fragments attributable to grass chaff were common elements, although it remains arguable whether these derive from thatching or crop processing. Cotton seeds are relatively rarer here, resembling the situation reported for Tumbe and Kimimba more than for the later sites. While wild and weedy seeds are as few here as at the other sites, Mazrui Tumbe contributes several new taxa to the record.

Having reviewed the primary archaeobotanical data for the four Pemban archaeological sites in question, the discussion now turns to a wider comparison of how these data accord with the model of urban food production proposed in Chapters 2 and 3. This is the subject of the next chapter, Chapter 7.

CHAPTER 7: TESTING THE URBAN FOOD PRODUCTION MODEL: ARCHAEOBOTANICAL PATTERNS ON PEMBA ISLAND

Introduction

In this chapter I test whether the Pemban macrobotanical record reflects urban food production patterns, as set out in Chapter 2 and applied to the Swahili case in Chapter 3. First, I briefly review the expected urban food production pattern, then I move to testing each expectation of the model in turn.

Expected Patterns of Urban Food Production: Adapting the Model to Pemban Grains

In chapter 2 I reviewed the most current models in urban food production, much of it drawn from Zeder's (Zeder 1991, 1999) study of urban animal remains from the Near Eastern site of Tal e Malyan in Iran. Zeder predicted that faunal remains could be used to distinguish between those with direct and indirect access to animal food resources, and ultimately to reveal the food production patterns under control of the emerging city and state. With respect to plant food remains, archaeobotanists have addressed the same dichotomy using a food producer/consumer framework (Hillman 1984; M. Jones 1985; van der Veen 1991) built through ethnographic observations of traditional crop processing activities and their material correlates (Dennell 1974; Hillman 1973, 1981; G. Jones 1984). Recent critiques of the various iterations of the producer/consumer model have served to question its broad applicability, particularly in non-triticoid (not growing wheat or barley) agricultural systems (Smith 2001), but have spurred a fresh generation of scholars to introduce new models based on different plant taxa (Reddy 1997), cultural contexts (D'Andrea et al. 1999; D'Andrea and Mitiku Haile 2002), research methods (Harvey and Fuller 2005) and interpretation of labor investment (Stevens 2003).

Generally, urban food production patterns are expected to differ from those in neighboring village communities in several ways. First, because urban centers are expected to be net importers of grain from the surrounding countryside, they will possess less evidence for grain cultivation or early-stage processing (harvesting, threshing) than village contexts. This can be measured using several factors, such as grain to chaff ratio, grain to weed ratio, and the proportion of large to small weed seeds (details below). Second, urban sites are expected to have grain stores in central locations under administrative control, and may also have some communal processing done in urban settings as well (Zeder 1991). This differs from village storage patterns, which will likely be at the domestic level. Third, because urban administrators require standardized food products to use as "currency" in trade or in supporting members of the community, it is expected that farmers will cultivate a narrower range of crops, typically those that produce high yield or those that are favored by trade partners. Thus, while farmers may traditionally prefer a range of cultivars to ensure an adequate food supply during good years and bad (e.g., Butler and D'Andrea 2000; Colson 1979), the demands of the urban economy for a uniform product will lower species diversity, particularly in urban settings.

Adapting the Urban Food Production Model to Swahili Cultigens

In order to best model urban food production patterns in the Swahili world, we must take the most important and well-tested aspects of the direct/indirect access and producer/consumer models and alter these to fit the agricultural systems expected for rice and African millets. Three aspects of rice and African grain processing differ from that of the triticoid grains, and these are fundamental to understanding the processing waste patterns that likely occurred on Pemba.

1. First, both rice and African grains can be harvested readily by hand using a knife. This eliminates much, if not all, of the weed seeds implicated in differentiating producers from consumers in models that use animal-drawn plows or hand-held scythes, in which sections of a field, including crops and weeds alike, are cut simultaneously (Dennell 1976; Hillman 1981, 1984; Jones 1985; Stevens 2003; van der Veen 1991). It also makes actual cultivation of grains difficult to verify (Smith 2001). Culm elements are the only other major by-product of harvesting, and these may be fed to animals or included in the stored or transported grain assemblage. Both of these scenarios eliminate straw from the archaeological record. Moreover, grass straw is best differentiated taxonomically using phytolith analysis (Harvey and Fuller 2005), which requires different methods of sampling, processing, and identification than the macrobotanical analyses performed by most archaeobotanists.
2. The second factor complicating the interpretation of rice and millet production pathways is the possibility that the grains may have been stored or transported while still on the head. This is particularly a concern with pearl millet, whose dense spike makes an efficient unit of trade or storage. All processing of grains transported on the head would then take place at its destination, the consumer site, either before storage or just prior to food preparation. In this situation, sites where grains are received on the head would show a remarkably similar pattern to sites where grains were both produced and consumed.
3. Rice grains and African millets do not regularly get parched as a pre-storage treatment, an action that in many wheat and barley systems contributes grains to the archaeological record. This means that African millets are exposed less frequently

and regularly to fire than predicted using the original producer/consumer models. This may lessen preservation in the archaeobotanical record.

As discussed in Chapter 2, archaeobotanists rely on crop processing models formed from ethnographic observations of non-mechanized farming activities. Typically this is done to form a relational analogy between the past and the present, as opposed to a historical analogy, which implies continuity of practice and culture (Wylie 1985). Relational analogy works on the assumption that materials and forms constrain the naturally wide range of human behaviors to fit a specific action and its end-products. For example, because pearl millet takes the form of a tall, single stalked plant with a compact spike, it requires a different set of processing behaviors than, for example, *mbungo*, the rounded globose fruits borne by lianes of *Saba florida* in Pemba's forests. Among the Swahili, non-mechanized farming practices observed today *may* closely resemble those used in ancient agriculture (Wright 1993); more importantly, however, the archaeobotanical record bears witness to the fact that similar crops were employed, as was the household-based production unit in operation today. This supports the validity of using ethnographic observations of Swahili plant food processing today to construct models of ancient agricultural practices. Readers should note that these models do not purport to reconstruct ancient behaviors exactly; rather they provide basic guidelines for understanding and interpreting ancient archaeobotanical assemblages. Following the ethnographic model provided for pearl millet by Reddy (1997), and by my own observations of contemporary processing of rice and sorghum by hand on Pemba, rice crop processing stages are noted in Figure 7.1 below, along with expectations of products and by-products at each stage.

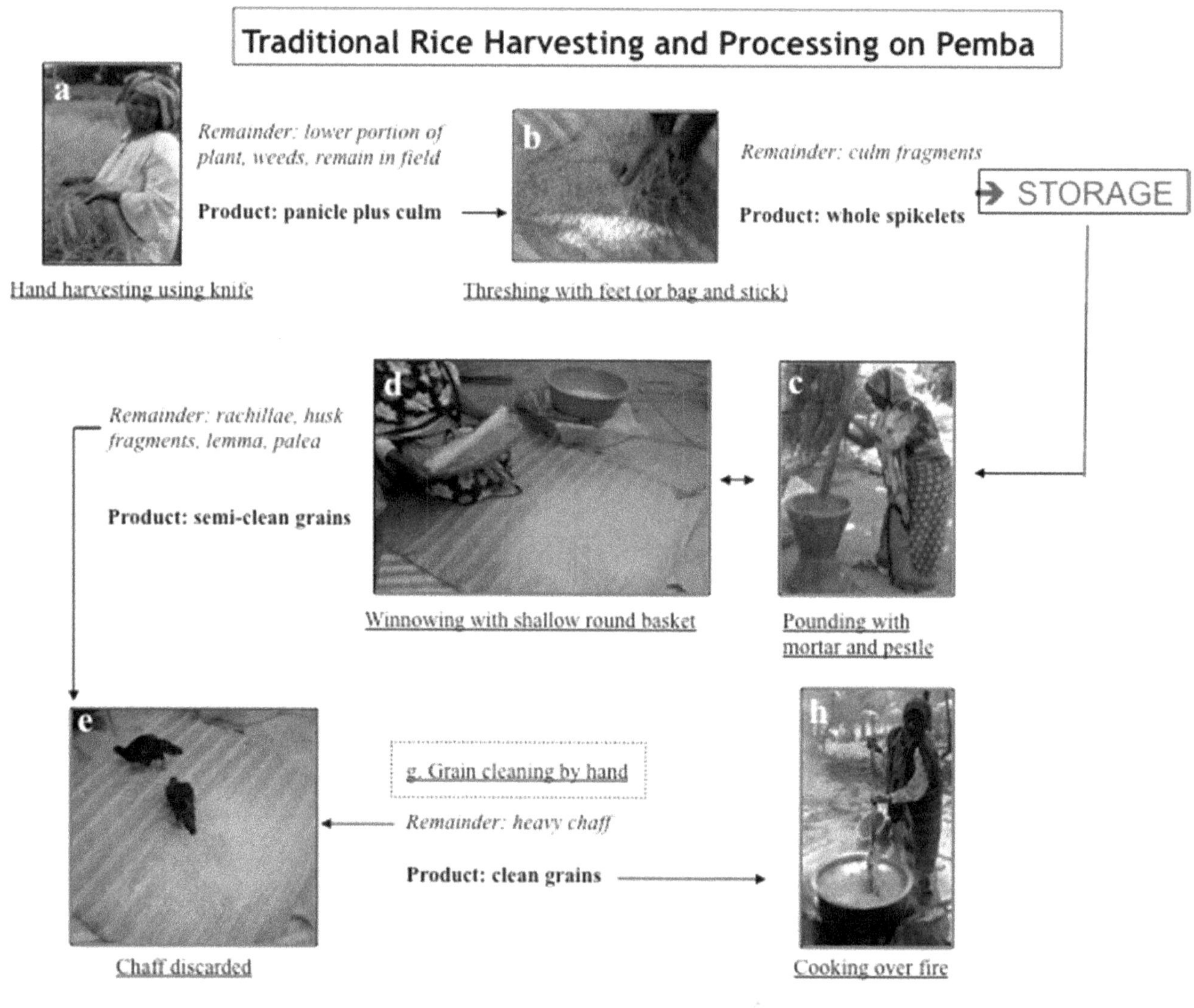

Figure 7.1. Traditional rice processing model showing products (in bold) and by-products (in italics) of each stage, based on ethnographic observations on Pemba made by the author, supplemented by Reddy (1997) and Harvey and Fuller (2005); (a) harvesting, (b) threshing, (c) pounding, (d) winnowing, (e) discarding residue, (g) cleaning, (h) meal preparation.

Testing the Urban Food Production Model

Measure 1: Grain to Chaff Ratio.
As discussed above and in Chapter 2, urban sites are expected to be net consumers of grains, while the surrounding countryside likely serves as the net producer responsible for the urban food supply. In the absence of telling weed seed assemblages, it is expected that the grain:chaff ratio may be the most revealing measure of a producer (low) or consumer (high) site. A value above one reveals that more grain than chaff was recovered; a value below one (but above zero) indicates a dominance of chaff. The data for the Pemban archaeological sites are presented in Table 7.1.

Rice. Following the data shown for rice, it can be seen that more chaff than grains were found across all sites and contexts except for Operation 2 at Chwaka. Considering that the most common chaff element found, the rachilla, is nearly as robust and identifiable as the grain itself, this can be interpreted as a genuine abundance of rachillae compared to grain. Moreover, as reported in Chapters 5 and 6, rachillae were more frequently found at sites than grains, thus rachillae proved to be more abundant as well as more ubiquitous.

Chwaka Rice Storage. The vast majority of rice across all four sites comes from Chwaka Operation 2, specifically two levels within Unit B of this exposure. Chwaka Op-2 is a series of earth and thatch house structures built atop one another (LaViolette et al. 2004). It is located just north of the Friday mosque, a prime location with a strategic view overlooking Micheweni Bay. As reported in Chapter 6, a burned cotton storage feature was found toward the bottom of the sequence, dating to the eleventh to thirteenth centuries (A.D. cal 1030-1240). The levels containing the rice concentrations were above this cotton feature, likely dating to the thirteenth century A.D., although this awaits verification through analysis of imported pottery. The rice concentration cuts through two layers (Layers 6 and 7), but was not concentrated into a recognizable feature like the cotton. Rather, the rice was scattered across a 2x2m block of the structure, called Op-2B (see figure 7.2). Other levels within Op-2B also contain significant amounts of rice grains and chaff, although not at the scale observed in Layers 6 and 7, which contained almost 14,000 grains and more than 3900 rachillae from only 29 liters of original sediment. It should be noted that in these layers, due to the quantity of rice recovered, rice grains were only counted when they were at or above 2.00 mm in size; thus the data reported here are a minimum value.

Table 7.1. Grain to chaff ratios for the Pemban archaeological sites. Ratios in parentheses include possible grain counts. Chaff counts are largely spikelet bases (rachillae) with some rice husk material in Chwaka Op2. Where one or the other value is 0, the ratio sign is used.

Time Period (A.D.)	Site	Context	Grain:chaff RICE	Grain:chaff PEARL MILLET
		Entire site	.25	13.4 (21.4)
7th-10th	**Tumbe**	Unit 14	.50	14.4 (21.8)
century		Site minus U.14	.19	10 (12.7)
	Kimimba	Entire site	0:0	1:0
		Entire site	2.83	10:0 (20:0)
11th-16th				sorghum = 5
century	**Chwaka**	Chwaka Op-2	3.57	1:0, sorghum = 3:0
		Site minus Op-2	.09	9:0, sorghum = 2:0
14th-16th		Entire site	.27	4:0 (4)
century	**Kaliwa**	Unit 5	.53	3:0
		Unit 4	.19	0:0
19th century	**Mazrui Tumbe**		.43	5:0 (10:0)

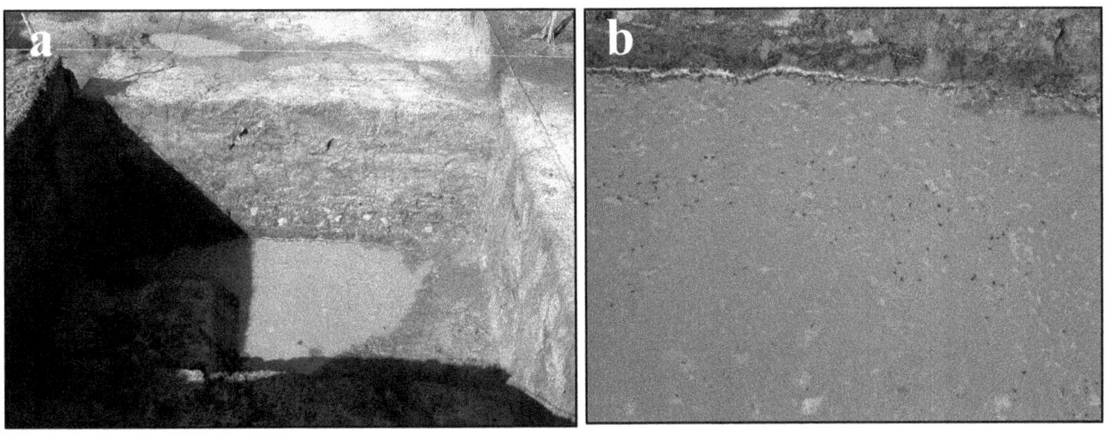

Figure 7.2. Chwaka Op-2B showing rice concentration; (a) unit wall looking west (wall is 2 m long), (b) rice grains floating in unit after a rainstorm.

Beside the rice concentration in the adjacent unit were found two cooking features, clay ovens called *mofa* (see Figure 7.3). While these have been found as relatively common features of stonetown houses (Horton 1996), these are possibly the only two found at the sites under consideration here, although Fleisher uncovered a *mofa* during test excavations at the Period 1 (A.D.700-1000) town of Bandarikuu on Pemba (Fleisher 2003:174, 177). These clay ovens could have had multiple culinary functions, such as the cooking of millets and rice, particularly in larger quantities, but Horton (1996:46) associates *mofa* specifically with the baking of unleavened bread prepared from millet flour. Taken together, the structural and feature data from Op-2B and Op-2C are consistent with a household storage unit and adjacent, likely outdoor, cooking area respectively. This will be discussed further below in the section dealing with possible grain storage facilities.

Pearl Millet. The grain:chaff ratio for pearl millet (and, where reported, sorghum), follows a different pattern from that just discussed for rice. Pearl millet grains were much more common than chaff elements. Of all the chaff elements, pearl millet spikelet bases were the most common, although one sorghum glume and one unidentified millet glume were also recovered. One possible complicating factor in the interpretation of the African millet ratio data is that the chaff components, particularly the pearl millet spikelet bases, are relatively more fragile and less easily distinguished compared to the grains themselves. This is not a major factor in the case of the sorghum glume, however, which was robust.

Figure 7.3. Probable clay oven (*mofa*) features in Chwaka Op-2C, 2 m south of rice concentrations shown in Figure 7.2 above. Scale bars are 5cm.

Considering ubiquity alongside frequency reveals that African millet chaff was both rare and sparse compared to the grains. In fact, pearl millet chaff was only recovered from Tumbe, and the sorghum glume from Chwaka. Thus, while African millet grains were found at all the sites (although in small quantities), chaff elements were found in two of the four sites. This differs from the pattern of rice grain and chaff distribution, where rachillae were more frequent and abundant than grains everywhere except the rice storage context in Op-2B.

Conclusions. Based on grain:chaff measures alone, a pattern emerges that splits along taxonomic rather than contextual or temporal lines. Across all sites, pearl millet grains are much more common than chaff, yielding a higher ratio than observed anywhere for rice. This is striking, considering that even when transported on the spike, chaff elements would be expected as part of the routine processing done for meal preparation. Even considering the relative fragility of pearl millet spikelet bases, this suggests that pearl millet may have been available in an already-processed state in some cases, although the presence of chaff in Unit 14 at Tumbe signals processing nearby.

Rice, on the other hand, shows a dominance of chaff across all contexts except for a storage context in Op-2B, which still showed an abundance of rachillae. Rice rachillae are found more frequently and abundantly across almost all contexts, signaling daily post-storage processing for meal preparation. Notably, the grain to chaff ratio for rice is lower at the stonetown (Chwaka) than at the nearby village (Kaliwa), which is the opposite situation of that predicted for urban sites obtaining grain from the surrounding countryside. Tumbe, Kaliwa, and Mazrui Tumbe all exhibit similar grain to chaff ratios, ranging between .25 and .43, while Kimimba contained no rice elements whatsoever.

Measure 2: Weed Assemblages

Weeds form a significant part of most agriculturally influenced archaeobotanical assemblages around the world. By all accounts, however, weed seeds are found only infrequently among all the Pemban archaeobotanical samples, and this has largely been attributed to the exclusion of weeds during the harvest using a hand knife to selectively cut crops. But the fact that weed seeds are found at all demands an explanation. A closer analysis of the weed taxa present and their abundance relative to grains is instructive.

The specimens listed below in Table 7.3 are considered small weed seeds, predominantly found below 1.00 mm. This is interesting considering that small, light weed seeds are removed during the early stages of crop processing. While many of the weeds could be identified to family, species is more difficult to determine due to fragmentation of the specimens and a lack of botanical sampling in eastern Africa. Generally, the families represented here include cosmopolitan weeds with wide geographic distribution today (Holm 1977). Readers should note that several "weed" taxa have good potential as vegetable crops, such as sorrel (*Rumex acetosa*), amaranth, and purselane (Grubben and Denton 2004), but they are included in this discussion of weeds because of their weedy nature (prevalence in disturbed habitats, difficult to eradicate, association with human habitation or cultivation) and therefore their possible connection with Pemban fields and gardens (Anderson 1997[1952]; Holm 1977; Koenders 1992).

Table 7.2. Weed seed distribution at Pemban archaeological sites (T=Tumbe, Km=Kimimba, Chw=Chwaka, Ka = Kaliwa, MT = Mazrui Tumbe)

Weed (*Species*, Family)	T	Km	Chw	Ka	MT	Total
Dock (*Rumex*, Polygonaceae)	10	2	7		4	23
Pink family (Caryophyllaceae)			5	3		8
Sedge/buckwheat family (Cyperus/Polygonaceae)	1		3	2	1	7
Legume family (Fabaceae)	1	1	1	1	1	5
Carpetweed (*Mollugo*, Molluginaceae)				1	3	4
Purselane family (Portulacaceae)	1		1		1	3
Small grasses (Gramineae)	1			1	1	3
UNID A (poppy family, Papaveraceae)	1	1			1	3
Amaranth family (Amaranthaceae)		1				1
Mallow family (Malvaceae)				1		1
Spurge family (Euphorbiaceae)	1					1
Goosefoot family (Chenopodiaceae)			1			1
Buttercup family (Ranunculaceae)			1			1
Spiderwort family (Commelinaceae)					1	1
Total weeds at site	**16**	**5**	**19**	**9**	**13**	**62**
Total weeds/L sediment	.061/L	.021/L	.056/L	.08/L	.13/L	
Grain: weed	19.4	3.2	833.2	6.6	1.9	

The weed assemblage across all sites is strikingly sparse. Only 62 weed seeds were identifiable to family, although undoubtedly more weed seed fragments were placed in the Unidentified Seed category because they were too fragmentary to be recognized. This weed seed count is low, particularly when compared with the volume of sediment floated at each site. Moreover, all the seeds under consideration are small and light, and as such should have been removed during the early stages of crop processing, such as those associated with post-harvest threshing and winnowing. As discussed above, weed seed elimination is high during the early stages of rice and African grain processing due to the selectivity of hand-harvesting. The low abundance of weed seeds may reflect this practice, in which case no weed seeds would be expected at all. Conversely, the weedy taxa above may represent plants growing close to households, either along pathways or in gardens. Seeds of these plants may have been swept up and disposed of in the fire. And, as discussed above, several of the weed taxa may in fact represent food plants, especially weedy greens, and thus may not have been implicated in crop processing activities.

A new perspective can be achieved through examination of the grain to weed ratio, where cultivated seed number is divided by the total weed seed number at each site. Producer/consumer models predict that producer sites will have a much lower ratio of grains to weeds, while consumer sites will have many more grains relative to weeds. . However, cultigen seeds outnumber weed seeds at all of the sites. This measure sets Tumbe and Chwaka apart considerably (particularly Chwaka) from the other two sites, but this is due to the large number of cultigen seeds found at the two larger sites Thus, the ratio of grains to weeds is less instructive in assessing the nature of Swahili food production on Pemba due to the scarcity of weed seeds across all of the sites under consideration.

Measure 3: Species Diversity

Under administrative control, urban food production is expected to yield increasing numbers of fewer taxa, because the administration specializes in a few subsistence or cash crops for the purposes of standardization in goods for trade or wage distribution. While the Swahili economy could conceivably demonstrate standardization in several categories (wood, fruit/nut), discussion here will be limited to seed taxa due to identification problems in the other botanical categories. A standard measure of diversity in ecological contexts is the Shannon Index described in Chapter 4. A lower value signals less diversity, either through narrowing of the taxa or dominance of one category.

Table 7.3. Shannon Diversity Index for Pemban archaeological sites.

Time Period	Site	Context	Seed Taxa	Shannon Index
7th-10th century	Tumbe	Entire site	16	1.30
		Unit 14	7	.56
		Site minus U. 14	14	2.59
	Kimimba	Entire site	14	3.40
11th-15th century	Chwaka	Entire site	18	.26
		Op-2B	9	.05
		Site minus Op-2B	16	1.19
14th-15th century	Kaliwa	Entire site	11	1.52
19th century	Mazrui Tumbe	Entire site	13	2.76

The diversity indices shown above signal the presence or absence of taxonomic concentrations at the different sites, and not the number of different taxa present. This is, therefore, a measure equivalent to unevenness. Indeed, Chwaka has the largest number of seed types yet the lowest seed diversity measure. The low diversity indices for both Tumbe and Chwaka are largely owing to a seed concentration at each site that overwhelms the diversity signal from the remaining contexts. At Tumbe, this is Unit 14, which yielded an abundance of pearl millet grains and most of the spikelet bases found at the site. At Chwaka, the low diversity is largely owing to the rice concentration found in Op-2B, although the cotton storage feature in Op-2C also contributed.

Both Tumbe and Chwaka stand out as centers of crop specialization in a countryside bearing a more diverse economic base. By the Shannon diversity measure, Kimimba is more than 2.5 times as diverse as Tumbe during the same time period, although when unit 14 is excluded from the calculation Tumbe's index approaches that of Kimimba. Similarly, Kaliwa is almost six times as diverse as Chwaka, but is only 1.3 times as diverse as Chwaka when Op-2B is excluded. By this measure, both Tumbe and Chwaka indicate that Swahili towns demonstrated increasing economic specialization through time, even if this was unevenly distributed throughout the site. Moreover, it is apparent that this specialization shifted through time, from pearl millet at Tumbe to rice, and secondarily cotton, at Chwaka.

Measure 4: Evenness of Distribution & Evidence of Storage

Urban food production is predicted to involve central storage under the control of the administration. This is particularly true of grain crops, which are harvested during a short period of time annually and require protection from post-harvest pest and mildew damage.

Swahili farmers today typically store grain for domestic use in containers within the house, under the bed for example, to monitor grain condition and to prevent rats from stealing food. While no storage structure per se was found at any of the sites excavated, two grain concentrations and one cotton concentration were observed during excavations, and subsequently verified through archaeobotanical analysis. These three concentrations will be discussed further here, as they are the major sources of the data that have thus far been used to differentiate Tumbe and Chwaka from their nearby villages.

Tumbe: Pearl Millet Concentration. Eighty-three percent of the pearl millet grains and 92% of the possible pearl millet grains found at Tumbe were located in one of two samples from Unit 14. Another small concentration was noted in Unit 56. All of these contexts originated in the lower levels of Tumbe, and Unit 56 provided the radiocarbon dated specimen with the early date for Tumbe (A.D. 620-700). These two contexts also account for 92% of all the rice spikelets recovered from Tumbe. Most of the pearl millet grains (76%) and spikelets (69%) at the site were recovered from a feature sample of 3 liters located in Unit 14, Layer 3, level 2 (Fea-7). This translates to a concentration of 70 grains/liter and 3 spikelets /liter for this feature, which is far higher than any other context except for the storage features discussed below at Chwaka.

While this context certainly appears to be a storage feature, it appears to be in a domestic context (LaViolette et al. 2003, Fleisher and LaViolette 2013). Unit 14 contained a high proportion of local pottery compared to imported pottery, and a moderate amount of worked iron and iron slag. Other evidence for metallurgy includes 3 tuyeres (clay pipes used in iron furnaces to draw in air from bellows) and a piece of copper, one of the few found at Tumbe. In addition to iron production, Unit 14 bears evidence of another local production activity as well: it contained 28 "bead-grinders", pottery sherds with linear grooves incised, possibly from bead making or other grinding needs. Interestingly, Unit 14 contained a considerable amount of shell, daub, and coral, which signals that while it had once been a house, it likely served as a midden after the house was abandoned.

A domestic storage context would be considered standard in the Swahili world, even in circumstances where elites were appropriating grain for trade products. Administrative structures are largely absent in Swahili archaeological sites, and public architecture such as mosques and tombs would not be appropriate storage facilities. What is unique about Feature 7 in Unit 14 is that grain storage appears to be unlinked to wealth in this instance. Unit 14 has a much lower imported:local pottery ratio than observed among other units at Tumbe, and shows more evidence of production than it does consumption. Thus, Unit 14 is best interpreted as a commoner household deposit, with Feature 7 representing the accidental burning of a storage unit during a house fire. This also possibly explains why other similar features were not found at Tumbe: the feature was created accidentally, then protected through midden deposition, creating an extraordinary preservation sequence.

Pearl millet was found all across the site of Tumbe, in relatively small amounts save for the context just described. Taken together, the evidence suggests that pearl millet was widely available to all members of the population, and larger quantities could be amassed by the average citizen for subsistence use or trade. In this case, Feature 7 does not appear to represent an administrative storage context, and it seems likely that other domestic storage features would have existed at Tumbe. Interestingly, the contemporaneous village site of Kimimba bears little evidence for cereal agriculture, containing only one possible pearl millet grain in 17 samples. Thus, while Tumbe does not appear to show the predicted urban food production pattern, the data suggest that Tumbe residents had access to more cultivated grains (pearl millet and some rice) and also coconut than their village neighbors at Kimimba. This indicates relative independence of the two communities in terms of plant food resources, which is not predicted by the urban food production model.

Chwaka: Rice Concentration. Ninety-two percent of the rice grains and 73% of the rice rachillae found at Chwaka were found within three composite flotation samples taken from Layers 6 and 7 at Chwaka Op-2B. This abundance of rice was not confined to a recognizable feature within the house, but rather was scattered across the entirety of the 2x2 m unit. Similar to the situation described above for Tumbe, the Chwaka rice concentration appears to be a domestic storage context that was charred in an accidental house fire, then covered over, in this instance by a new house floor, and thus protected from further decay or disturbance.

This likely storage feature is highly useful in reconstructing production patterns prior to storage. The high number of rachillae observed indicates that rice was likely stored with little pre-storage processing. This helps explain the ubiquity of rice rachillae across all samples from the site. Meal preparation in all households would require routine pounding and winnowing to separate the grain from the husk, and the discarded chaff would have a higher chance of entering a fire than the subsequently cooked grains. While the majority of rice was concentrated in these three samples, overall rice was well distributed across time and space at Chwaka. Rice rachillae were present in all archaeobotanical samples, and rice grains were observed in 88% of samples. We will return to this issue of pre-storage processing in the next section of this chapter, as it informs our discussion of labor in agricultural strategies.

Chwaka: Cotton Concentration. In addition to rice, cotton was a common element of the Chwaka archaeobotanical record. Cotton was present in 88% of samples, with almost 600 cotton seeds estimated across the entire site. Much (68%; 404 seeds estimated from 6.43 g) of the cotton at Chwaka was located in a feature sample from one of the lowest layers of Chwaka Op-2C (Fea-192, Layer 12, Level 1). This feature contained little artifactual material (some local pottery, one glass bead), but was archaeobotanically very rich. Many whole cotton seeds were recovered (see Figure 7.4), and only cotton seeds and fragments above 2 mm were pulled; therefore the 404 estimated seeds represents a minimum for this sample. In addition to the cotton, an impressive amount of wood, fruit/nut fragments and rice chaff was recovered. The rice chaff was represented by all of the spikelet elements, including 24 spikelets (grain removed), 327 rachillae, 40 possible rachillae, 53 fragments of palea/lemma, and 79 embyros. This is an impressive array of chaff materials; whole spikelets were relatively rare, and palea and lemma pieces were exceedingly rare, across the entire site. Very few rice grains (16) accompanied this abundance of rice chaff in the cotton feature. Considering the large quantity of wood and fruit/nut tissue discovered in this feature, as well as the low grain: chaff ratio (.03), this feature is consistent with a hearth located close to domestic production activities, including the removal of rice from the spikelet and cleaning the cotton lint off the seed to spin yarn. Spindle whorls were found in the layers directly above and below this feature, providing further evidence that this was a locus of production within the house.

Figure 7.4. Cotton recovered from Feature 192 in Chwaka Op-2C. Scale is in mm.

Processing Labor and Social Relations on Pemba
As discussed in detail in Chapter 2, Stevens (2003) used elements of the producer/consumer model to interpret social organization by assessing labor mobilization strategies during different crop production phases (see Table 7.4). Stevens predicts that communities where household labor could join forces in communal processing activities would be able to clean grain more effectively prior to storage, resulting in a higher proportion of cleaned grains within a storage context and a weed assemblage with a higher proportion of large or heavy weeds.

Conversely, if individual fields were harvested using only household members or extended kin as labor, then the harvested grain would undergo only minimal processing. As a result, batches of grain would be stored in spikelet or husked form, and would contain many small weeds in addition to large ones. This difference in stored grain form would also be apparent in the routine daily processing of stores required for meals. The relatively cleaner grain produced by communal efforts would result in less and different waste products during routine processing than would the household stores with relatively more chaff and weeds produced by less intensive pre-storage production.

All the evidence presented previously in this chapter allies the Pemba archaeobotanical data most closely with Stevens' household labor model. There are, however, several problems with the application of this model to the Pemba data. First, Stevens' model was conceived for use with triticoid grasses, namely the wheat and barley grown in Roman Britain from which Stevens takes his examples. As demonstrated above, both rice and African grain crops do not conform to the expectations of crop processing based on wheat and barley, particularly where weed assemblages are concerned. Considering that much of Stevens' interpretation rests upon the analysis of weed seeds, the relatively weed-poor archaeobotanical record on Pemba cannot be tested directly against Stevens' predictions.

Table 7.4. Stevens' (2003) model of the archaeobotanical interpretation of post-harvest pre-storage labor mobilization for crop processing, adapted from Figure 7 and pp. 72-73.

Variable	Household labor	Communal labor
Grain: weed	Low	High
Proportion of large weeds	Low-moderate	High
Grain: chaff	Low-moderate	High

There are, however, several facets of Stevens' argument that may be applicable to testing the connection between town and village on Pemba Island. Stevens' expectation that increased social organization may alter crop processing strategies can be applied to the urban food production model. If larger centers such as Tumbe and Chwaka were commanding grain from the countryside for their own trade and subsistence use, then it is reasonable to expect that they could command labor forces to harvest the grain, and potentially process grain with increased labor compared with household farmers. Thus, regional centers should store relatively cleaner grain, and the routine meal processing of grain stores should yield less chaff than households in the surrounding villages. As shown in Table 7.1 above, the routine processing signatures of Tumbe and Chwaka (i.e., the non-storage feature contexts) are very similar to those observed in the village of Kaliwa and the Mazrui Tumbe settlement; in fact the average grain to chaff ratio of the larger sites is *lower* than the village sites, a pattern not predicted by the urban food production model.

If, as I argue, the Chwaka storage concentration in Operation-2B can be considered typical at Chwaka, then the prevalence of chaff elements there, particularly rachillae, signals a low level of pre-storage crop processing taking place. This suggests a household level of rice production in operation at Chwaka, and indeed across all the sites at which rice is documented. This implies that larger centers on Pemba were not reliant upon, nor could they commad the labor of, rural communities for food production. This system of household food production will be summarized in the following section of this chapter. In the final chapter (Chapter 9), I discuss the implications of a domestic food production practice on Pemba, and in Swahili economy and politics in general.

A Household Model of Swahili Food Production

The bulk of the evidence provided in this chapter fails to support a model of urban food production, characterized by rural settlements producing grain for urban consumption, existing on Pemba during the occupation of the four sites in question, especially after A.D. 1000 when Chwaka stonetown is settled. Instead, the archaeobotanical data support a model in which individual households were the main unit of production. At Tumbe and Kimimba, the seventh to tenth century sites, this is seen in the relative economic independence of the two sites: Tumbe shows more evidence of cereal agriculture and introduced foods (e.g., pearl millet and coconut), while Kimimba largely lacks grain and has an abundance of local fruit and nut taxa and some well preserved root/tuber specimens. By the time of Chwaka (eleventh to sixteenth centuries) and Kaliwa (fourteenth to fifteenth centuries), both large and small sites appear to rely on rice agriculture, and the vast majority of contexts show evidence for post-storage processing, indicating a low labor investment in pre-storage processing not consistent with a standard model of urban food production.

While it is virtually impossible using macrobotanical evidence to prove that the plants under consideration were actually grown on Pemba (Harvey and Fuller 2005; Smith 2001), local household ownership of fields and gardens is the most parsimonious explanation for the wide distribution of plant remains among the observed villages and towns, and accords with the ethnohistoric record for the Swahili coast (Horton and Middleton: 133-134). Thus, families would have planted, tended, and harvested their own plots, with occasional help from kin or hired help as necessary. Because of the narrow time frame of ripening and harvest, all available hands would have been conscripted to help with the harvesting of rice and pearl millet, as well as other grains such as sorghum that ripen in July and August. This may help explain the low levels of pre-storage processing. I hypothesize that a limited labor supply demanded streamlined processing in the field, while routine daily processing could afford to be more time-consuming due to comparatively lax time constraints. This is not to say that the daily household tasks of Swahili families were minimal – there are myriad chores involved in child-rearing, animal tending, fishing, cleaning, tending of gardens and fields, and craft production, in addition to meal preparation. Rather, I suggest that the time pressure at harvest was considerably stronger than that observed on a daily basis when households, possibly containing extended kin, had relatively more time and labor to assist in routine chores such as pounding and winnowing rice for meals.

As discussed in Chapter 2, the household mode of production is characterized by the household being the primary labor unit and production locus. This contrasts with the centralized food production model predicted for urban or state systems, where the administration controls some aspects of food production in order to obtain adequate materials for trade and local provisioning. One of the most significant ramifications of the household mode of production is that the household could potentially be the primary unit of trade. That is, any family that could produce surplus goods could potentially commandeer these for local or foreign exchange, unless prevented from doing so by cultural norms or limited trade networks. This is a key departure from the centralized food production model: because the administration controls the surplus yielded by production, it benefits the most from the mobilization of surplus in trade relations and site or staff provisioning. In the centralized food production model, rural food producers are largely divorced from the benefits or participation in larger spheres of local and trade economies. In the next chapter, I engage in a broader discussion of how the plant food economies in northern Pemba Island intersect with other local production and consumption economies.

CHAPTER 8: HOUSEHOLD PRODUCTION AND CONSUMPTION IN LOCAL AND REGIONAL CONTEXT.

Introduction

In this chapter, I integrate the archaeobotanical data with the faunal data from Pemban sites, as well as the evidence for other arenas of local production, to determine if the proposed mechanisms of plant food production are consistent with other local production domains. Animal food production will be discussed first, followed by craft industries evident in local pottery, iron, spindle whorls, and grinders. The distribution of production activities is then compared with patterns of material wealth found among Swahili towns and villages.

Animal Economies: the Shell and Faunal Records in Northern Pemba

As discussed in Chapter 3, faunal investigations at several Swahili archaeological sites have illuminated our understanding of the range of domesticated animals, wild game, fish, and shellfish consumed (Horton and Mudida 1993; Mudida 1996; Wilson and Omar 1997). Recent surveys and excavations on northern Pemba Island have revealed interesting patterns in terrestrial and marine animal resources across time and space that inform the present discussion of food production pattern (Cain 2005; Cain in LaViolette et al. 2003; Msemwa and Mwebi in Fleisher 2003). It should be noted that site-dependent differences in faunal preservation limit our ability to compare directly quantitative data across the sites, but trends in relative taxonomic representation are real and meaningful (Cain 2005).

Shellfish

Shells of mangrove and shallow bay mollusks are common elements of the northern Pemban archaeological record. The close proximity of the archaeological sites to mangrove swamps and Micheweni Bay suggests that shellfish could be collected expediently, with few, if any, tools. Shell remains are most likely indicative of shellfish consumption activities, but possibly also shell bead production, although the latter activity would be expected to yield a more fragmented assemblage, possibly even pulverizing the shells in the process (Mudida 1996). Due to the fact that shellfishing yields relatively little protein for the collecting effort expended, shellfish are generally considered a lower-quality food, one restricted to poor or otherwise marginalized members of a community, as observed today (Fleisher 2003:354-55, contra Cain 2005).

Patterns of shellfish consumption noted for Kaliwa and Chwaka show a different reliance upon these marine resources during the same period that may be tied to urbanization (Fleisher 2003:363; Msemwa in Fleisher 2003). While shellfish remains were found throughout all time periods at Chwaka, they are most abundant during its growth (A.D. 1000-1300), and wane considerably at the height of this stonetown's occupation (A.D. 1300-1500). In contrast Kaliwa's (A.D. 1300-1500) faunal record is dominated by shellfish. Fleisher (2003:363) takes this as a sign of economic independence between this wealthy stonetown and its neighboring, relatively impoverished, village, observing that "[i]f we can use density of shell material as a proxy for shellfish consumption, then the residents of Kaliwa ...were consuming more than ten times the amount of shellfish than their contemporaries at Chwaka, 2 km away." Furthermore, Cain (2005) notes that the Chwaka shellfish assemblage contains a larger proportion of bigger, high-return taxa than that of Kaliwa. Interestingly, shellfish dependence also drops dramatically at Shanga after urbanization (Horton and Mudida 1993; Mudida 1996), adding further weight to a hypothesized correlation between access to wealth and the proportion of shellfish in the diet.

Fish

Fish remains, especially vertebrae, were also common at Chwaka and Kaliwa. Poor preservation conditions at Kimimba, and ongoing analysis of the 2002 and 2004 fish remains, precludes me from including information about the first millennium Pemban sites here. Faunal samples from units excavated by Fleisher in 1999 at Kaliwa and Chwaka were analyzed by Otego Mwebi at the National Museums of Kenya (Mwebi 2000 in Fleisher 2003:366). By identifiable bone count (NISP), fish bones represented 88% of the faunal remains from Kaliwa but only 29% of the faunal assemblage from Chwaka (see Table 8.1). Similar to the shellfish exploitation patterns noted above, the majority of fish recovered archaeologically thrive in the shallow saltwater habitats located near Chwaka and Kaliwa, yet some major differences in taxonomic representation exist between the two sites. The presence of specialized fish technologies at Kaliwa may account for the greater abundance of deepwater and poisonous fish taxa there as opposed to Chwaka. For example, a fish species likely obtained through off-shore fish trapping, *Naso hexacanthus* (sleek unicornfish), accounted for 24% of identified fish bones at Kaliwa, while this taxon represented only 6% of the Chwakan fish assemblage (Fleisher 2003:373-374). Furthermore, poisonous fish taxa (e.g., Tetradontidae, which are deadly unless specially prepared) are much rarer at Chwaka (.1% of all identifiable faunal taxa) than at Kaliwa (.9%). Taken together, the fish data suggest that Kaliwans were much more dependent upon fish than residents of Chwaka, and that people at Kaliwa were trained in special fishing and preparation techniques to allow greater marine exploitation. The use of poisonous fish in particular may indicate that Kaliwa experienced food stresses that Chwaka was more sheltered from. The next section addresses the non-fish components of the faunal record to examine why and how Chwakans may have been buffered from food scarcity.

Table 8.1. Faunal remains other than fish from Chwaka and Kaliwa indicated by percent NISP (from Mwebi 2000 in Fleisher 2003 Tables 8.10, 8.11)

Common name (*Scientific name*)	Kaliwa	Chwaka
Domestic Cow (*Bos taurus/indicus*)	3.6%	23.2%
UNID large mammal (Mammal IV)	5.8%	38.4%
Domestic Sheep/goat (*Ovis/Capra*) plus UNID medium mammal (Mammal II)	.1%	.8%
Domestic Chicken (*Gallus gallus*)	.9%	3.9%
UNID bird	-	.1%
Dugong, sea cow (*Dugong dugon*)	.6%	.4%
East African black mud turtle (*Pelusios subniger*)	<.1%	-
Green Seaturtle (*Chelonia mydas*; *Chelonia* sp.)	.7%	2.5%
Domestic Cat (*Felis domesticus*)	<.1%	1.3%
Black Rat (*Rattus rattus*)	.1%	.1%
Mangrove Crab (*Scylla serrata*)	-	.1%
Total Non-Fish	**11.9%**	**70.7%**
Total Fish	**88.1%**	**29.3%**
Total Fauna	**100%**	**100%**
Total Number of Identified Specimens (NISP)	**2282**	**1938**

Non-Fish Fauna (Mammals, Birds, Reptiles, and Crustaceans)

As expected from above, non-fish taxa were dominant at Chwaka, comprising almost 71% of identifiable elements (see Table 8.1 below). In fact, all non-fish taxa were more abundant at Chwaka than at Kaliwa, except for the dugong, which is caught in shallow saltwater bays that support seagrass growth. The exploitation of dugong is consistent with the maritime focus of Kaliwa's subsistence economy. It should be noted, however, that the higher species diversity and abundance observed among the non-fish fauna at Chwaka may be an artifact of the better bone preservation conditions existing there (Cain 2005).

Domestic cattle and unidentified large mammal bones dominated the non-fish faunal record, a trend corroborated by preliminary analyses of the 2002 and 2004 field seasons (Cain 2005). This pattern of cattle dominance among non-fish fauna persists at both sites, although clearly the cattle at Chwaka contributed more significantly to the overall diet. Chwaka displays a slightly more diverse array of non-fish fauna, with relatively more dependence upon domesticates than Kaliwa – ovicaprids and chicken are better represented at Chwaka. Cattle also dominate the non-fish bone assemblage at Tumbe, the seventh to tenth century settlement preceding Chwaka. Here, cattle account for almost 90% of the non-fish fauna (Cain 2005), a value comparable to both Kaliwa and Chwaka when the Mammal IV category (comprising large mammals) is included.

Summary

Taken together, the shell and faunal records of Kaliwa and Chwaka indicate rather different subsistence economies taking place at these two sites. In general Kaliwa is specialized towards marine resources, including shell, fish, turtle, and dugong, while Chwaka bears evidence of a predominantly domesticated meat supply. These patterns are not static, however, particularly with respect to the roles of cattle, chicken, and sea turtle through time. At Kaliwa, these three elements of the non-fish record increase through time, while overall fish abundance decreases by almost 20% over the two centuries of Kaliwa's occupation (Fleisher 2003:379 Table 8.10). The longer sequence at Chwaka demonstrates an even more dramatic shift in resource use. At the initial settlement of Chwaka (A.D. 1000-1200), fish were the most common element (66%) of the faunal record, but a major shift took place in the twelfth century that reversed the roles of fish (now 27%) and domesticates (suddenly over 70%) within the zooarchaeological record. Between A.D. 1300 and 1500, domesticates increased only slightly, and fish dropped even further to 20% of the faunal assemblage. This rapid increase in the abundance of domesticates, the majority of which are cattle bones, in the Chwaka assemblage can be seen as a development parallel to urbanization. A strikingly similar pattern can be observed during the growth of Shanga, where fish dominate the faunal record until the 12th century until their abundance drops dramatically throughout the following three centuries (Horton and Mudida 1993; Mudida 1996). One interesting difference between Shanga and Chwaka is the hunting of wild game at Shanga, and the consistent lack of evidence of such activities at Chwaka. This may be attributed to the relative lack of wild mammals in general on Pemba, as noted in Chapter 3 and described elsewhere (Koenders 1992; Walsh 1995). The paucity of wild game may also reflect that Pembans were not as culturally connected with the coast and hinterland, where wild game was abundant and was traded to the Swahili from nearby communities (Horton and Middleton: 133-134).

What do the differences in animal food production at Kaliwa and Chwaka imply with respect to urbanization? Cain (2005) posits that the different animal exploitation patterns represent economic integration through food sharing between specialized production domains. In this scenario, Kaliwa would catch and possibly process the majority of the fish destined for consumption at Chwaka, and Chwaka's domesticates would provide both sites with non-fish meat. Evidence of such a relationship between the two sites would be most clearly indicated in the representation of body parts, because fish and, more likely, non-fish meat would have been butchered prior to transport and distribution (Lyman 1987). Mwebi (2000 reported in Fleisher 2003:375), however, notes that elements were relatively evenly distributed between Chwaka and Kaliwa, which lends support to the hypothesis that each site was largely responsible for killing, butchering, and consuming its own meat supply. Instead of economic integration, the parts distribution data may indicate that each site was relatively independent in terms of obtaining protein and that the stonetown of Chwaka was relatively better stocked with domesticates than the nearby village of Kaliwa, which turned to the sea for its meat sources.

The shell and faunal data retrieved thus far from the Pemban archaeological record have interesting intersections with the archaeobotanical data as presented in Chapters 5-7. The image of economic independence drawn from the non-fish faunal data at the later sites of Kaliwa and Chwaka matches well with archaeobotanical observations that Kaliwa apparently did not provision Chwaka. Additionally, the archaeobotanical evidence shows more similarity in the dominance of the domesticates rice and cotton at these two sites, although they differ in their relative abundance of coconut (higher at Chwaka) and other fruit/nut (Kaliwa), which may signal that Chwaka had more control over coconut resources. Thus, while both the archaeobotanical and zooarchaeological signatures of Chwaka and Kaliwa suggest economic independence, the botanical data show more similarities in resource use than do the shell and faunal remains.

Food Production in Context with Craft Production Practices

As noted in Chapter 3, Pemba's archaeological sites contain widespread evidence for several types of craft production (Fleisher 2003; LaViolette 2000; LaViolette et al. 2003; LaViolette et al. 2004). Notable among these are local pottery production, ironworking, cotton fiber and textile production, and the manufacture of beads. Trends through time in these craft industries shed light on how and why food production appears to be relatively independent at each site, particularly during urbanization.

Pottery

Locally produced earthenware ceramics are the most common element of the archaeological record on Pemba, as elsewhere along the Swahili coast. Local wares were hand-crafted using a combination of press molds, coils, and rings to form a variety of vessels decorated using combinations of ochre and graphite paints/burnishes, incised lines, and punctations (Fleisher 2003:232-234; Sinopoli in Wright 1984). While the Pemban ceramics show strong similarities with other regional ceramic assemblages in terms of decoration, demonstrating common cultural heritage with and/or communication among other coastal communities, a distinctive Pemban tradition was discerned by Fleisher in his recent and extensive analysis. This "constellation of particular decorative motifs that occur on Pemba" includes "the prevalence of punctuates below incised decorations" and a chevron design (Fleisher 2003:261; see Figure 8.1). Moreover, the diversity of decorations within early (before A.D. 1000) sites on Pemba indicates that pottery was produced by a variety of people for local use, while a significant decrease in decorative types after A.D. 1100, accompanied by a decrease in idiosyncratic variability, may signal the development of incipient pottery specialization and standardization (Fleisher 2003: 262-263).

Figure 8.1. Local Tana tradition pottery from Pemba Island. Photo courtesy of the Pemba Archaeological Project.

Change in vessel forms during the period of urbanization is striking. Swahili vessels fall into two generalized categories: restricted forms, in which the rim diameter is not the maximum body diameter (e.g., pots, jars); and unrestricted forms, where the rim represents the maximum vessel width (e.g., bowls, platters) (Chami 1994; Horton 1996). Jars and restricted bowls, which dominate the Pemban ceramic record prior to A.D. 1000, wane in popularity until they all but disappear after the twelfth century (Fleisher 2003:264-265; LaViolette et al. 2003). In their place, open bowls and wide-mouthed jars expand in use until they comprise up to 90% of any given ceramic assemblage after 1100. This shift towards more open vessels coincides directly with the emergence of a rice-based cuisine on Pemba, as presented in Chapters 5-7 of this thesis. The different requirements of cooking and sharing a meal made with rice, as opposed to the more liquid preparations involving millets, may have inspired an increase in open jar and bowl forms. While the shift from a jar-based assemblage to a bowl-based one seems to persist elsewhere along the coast (Chami 1994), only on Pemba do we now have archaeobotanical material to demonstrate a correlation between vessel form and diet. It will be interesting to determine if this connection exists elsewhere along the coast, evidenced in the ceramic and archaeobotanical records.

African millet preparation may leave distinctive wear patterns on ceramics (Reid and Young 2000), which may be a consideration for future analyses. For example, Reid and Young (2000) observed modern finger millet processing in Uganda that involved removal of the chaff through roasting (which blackens the pot's exterior) and vigorous stirring (which thins the pot's interior walls) in a rounded pot tipped to one side over a hearth. These exact patterns were noted in 48% to 83% of sherds among several vessel forms from seven Ntusi sites in the interlacustrine region of eastern Africa (eleventh to fifteenth centuries A.D.), and comprised up to 33% of the total ceramic assemblage.

Iron

Project excavations in 2002 and 2004 revealed more about iron-working activities than did Fleisher's 1999 survey data, hence the data cited here include all sites sampled archaeobotanically. The iron-working data here originate in the two unpublished interim reports by LaViolette et al. (2003, 2004). Iron production activities appear ubiquitous at the seventh to tenth century town of Tumbe: all test units dug there in 2002 contained iron slag, and 15 tuyere fragments (ceramic tubes that deliver air pumped by bellows) from iron furnaces were recovered. This pattern is echoed at the contemporaneous village of Kimimba, where fragments of iron slag and finished iron were found dispersed throughout the site. Similarly, Chwaka also bears evidence of widespread iron production during the eleventh to sixteenth centuries. Kaliwa bore evidence of potential furnace-type features in one unit, and archaeological survey of the surrounding area turned up a Tana tradition (seventh to ninth century; contemporaneous with Tumbe) site with impressive evidence of iron smelting in the form of abundant slag and a long slag-encrusted tuyeres pipe. Iron fragments and processing remains are currently under investigation by Bertram Mapunda of the University of Dar es Salaam, an iron-working specialist and co-director of the Pemba Archaeological Project, so detailed information about iron production evidence on Pemba is forthcoming. Yet the abundance and dispersed distribution of iron smelting products in preliminary investigations at Tumbe and Chwaka lend further support to Mapunda's (2002) hypothesis that individual households were the locus of iron production, through the use of small crucibular furnaces. This view of iron production as small-scale, dispersed, and localized within households matches the production mode described in previous chapters for agricultural production, and strengthens the hypothesis that production activities were not centralized or dominated by stonetowns in Pemba's past.

Spindle Whorls & Cotton Production

Spindles were used in spinning thread from fiber, although *which* fiber has been a subject for debate until the findings of cotton seeds in this study confirmed that cotton was among the fibers employed. While spindles were likely made using a variety of materials, only ceramic and stone spindle whorls have preserved archaeologically to signal spinning activities in the past. Ceramic whorls on Pemba were formed predominantly from broken pieces of pottery, imported as well as local, which suggests an affinity with northern Swahili material culture traditions at Shanga and Manda. This differs from finds at the Tanzanian coastal center of Kilwa, where whorls were newly formed from clay (Fleisher 2003:328). Archaeological spindle whorls are apparent only after A.D. 1000, both on Pemba and at the site of Shanga (Horton 1996). Considering that cotton seed fragments were found at both first millennium sites on Pemba (Tumbe and Kimimba, reported in Chapter 5), it may be reasonable to assume that spindles in these earlier communities were formed from organic materials such as wood, coconut shell, gourd rind (e.g., *Lagenaria siceraria*, the bottle gourd), large seeds, fruit stones/shell, or wax (Liu 1978). Cotton fiber was most likely spun to weave cotton cloth for local use, although it remains possible that other fibers (possibly coconut husk?) were spun for creating fish nets, as indicated by the presence of off-shore fish taxa.

Evidence for thread and cloth production, as signaled through spindle whorl frequencies, indicates that production activities were widespread among second millennium sites in northern Pemba (Fleisher 2003:300; LaViolette et al. 2003, 2004), but appear to have been absent at Tumbe (Fleisher and LaViolette 2013). For example, 32 spindle whorls were recovered from Chwaka during excavations in 2004, representing all excavation units except for the mosque, but were also found at the nearby village of Kaliwa (n=4), in midden and household deposits. As presented in Chapters 5-7 of this thesis, cotton finds from these same sites provide further evidence for the ubiquity of cotton production: cotton was present in 90% of samples from Chwaka, and 100% of samples from Kaliwa. This represents an increase in cotton production from the earlier sites of Tumbe and Kimimba, where cotton frequency was 24% and 30%, respectively. Thus, cotton production was not limited to certain sites (i.e., center or countryside), nor to specialized production contexts within sites. However, intensification of production is observed as a dramatic increase in cotton ubiquity after A.D. 1000.

Bead Grinders

Bead grinders are pottery sherds or, less often, pieces of ground stone, bearing linear grooves of regular dimensions, apparently signaling the revival of broken pottery for use as a grinding surface for items with a rounded edge (see Figure 8.2). In a study of over 600 grinders from Tumbe, Flexner (2005) determined that the semi-circular grooves ranged from .3 to 1.8 cm in width and .1 to .7 cm in depth. The most commonly proposed item produced by such grinding is shell beads, due to the congruency of the groove dimensions with shell bead dimensions. Paradoxically, however, neither shell beads nor production debris were found on Pemba, either in sediment screens (1/4 inch) or in the heavy fractions of flotation samples (1-2 mm). It could be that shell waste is pulverized beyond recognition, or that manufacturing activities took place away from habitation areas of the settlement. It has been suggested that shell beads were all produced for export, but that would make shell bead production unique among the other production spheres on Pemba, which seem oriented towards local household consumption. Horton (1996:327) suggested that perhaps bead grinders were used to make ceramic spindle whorls; however these two artifact classes have disparate chronological distributions, based upon excavations to date. Considering that we now have evidence for cotton before A.D. 1000, but lack evidence for contemporaneous spindle whorls, it remains possible that bead grinders were used to fashion organic (ie, wood, coconut shell) spindle whorls, in addition to other potential uses. I could not detect any microscopic plant remains (e.g., phytoliths from coconut or bottle gourd) in the grooves of several bead grinders submitted for analysis using both light and scanning electron microscopy. Until further evidence comes to light, it remains most likely that bead grinders were employed in the manufacture of shell beads destined for long-distance trade, or for some other purposes yet to be determined.

Figure 8.2. Selection of grinders from northern Pemba Island. From Fleisher and LaViolette 2013:117 Fig. 7.

Regardless of the function of bead grinders, their abundance and distribution among early sites is telling. More than 3,000 bead grinders were recovered from Tumbe units excavated in 2002 and 2004, of which the vast majority was formed from local pottery (98.6%). This represents the highest density of bead grinders at any known coastal site by far (LaViolette et al. 2003). Bead grinders were present in all units investigated, but were unevenly distributed: Units 63 and 64 contained many more bead grinders, sometimes several hundred in one 1 m x 1 m x 10 cm level. Unit 63 likely represents an abandoned house that was subsequently used as a midden, while Unit 64 contained midden fill (LaViolette et al. 2004), so possibly the concentration of bead grinders in these contexts represents household trash accumulated from neighboring houses. In contrast, 80 bead grinders, all from local wares, were recovered from Kimimba during survey and excavation. These were distributed throughout roughly half of the excavated units, although a concentration was detected in one midden context (Fleisher 2003:333; LaViolette et al. 2003). Thus, whatever activities are represented by bead grinders took place at both Tumbe and Kimimba, although such activities were more intensively carried out at Tumbe. Moreover, the distribution of bead grinders may not be linked with households, as "bead grinders are generally found in deposits with lower densities of daub…[and] tend to be concentrated in pits or in discrete levels" (Fleisher 2003:333). This indicates that bead grinding may have occurred off-site, or in less-inhabited areas of a site, with grinders being deposited in middens after their use life.

Summary

A survey of the different major craft activities at sites in northern Pemba reveals a general pattern of dispersed production, both between sites and among household within sites. While local potting and iron smelting practices are apparent in all time periods, bead grinders and durable spindle whorls show distinct, temporally restricted, distributions. All craft production domains exhibit a tendency towards decentralization, a pattern echoed in the food production practices discussed earlier in this chapter. Thus, it appears that the town of Tumbe and the later stonetown, Chwaka, were not in direct control of local production practices and the redistribution of local products in regional trade systems. Rather, the evidence to date suggests that both towns and villages could produce crafts, and that production activities were relatively small scale and distributed throughout the community, likely at the level of the household.

The predominance of household production among Swahili settlements in northern Pemba is a significant finding (Fleisher 2003; LaViolette et al. 2003, 2004). The opportunity to sample the production and consumption activities of communities outside stonetowns offers a unique perspective on Swahili urbanization, particularly with respect to the economic integration of a stonetown with its countryside (LaViolette and Fleisher 2005). Previous conceptualizations of Swahili urbanization have assumed a connection between wealth and power, both of which were identified with the stonetown, which obtained products from its postulated sphere of control in the local countryside (Horton and Middleton 2000; Kusimba 1999; Wright 1993). Yet if production itself was dispersed throughout communities, then it is reasonable to predict that each household that could produce a surplus could have access to wealth, regardless of whether it was located in a stonetown. Having documented the decentralized nature of production within and among communities, I now turn to a discussion of patterns of consumption among Swahili households, focusing on the abundance and distribution of items associated with wealth, including imported pottery and glass.

Distribution of Wealth in Northern Pemba

In addition to the abundance of locally produced artifacts, Swahili sites often contain imported goods, including pottery, glassware, stone beads, copper artifacts, and steatite (chlorite schist), among others. These imports were obtained through Indian Ocean trade with peoples in Madagascar, the Arabian peninsula, and South Asia (Horton 1996:271). The presence of these items archaeologically is associated with the wealth needed to secure them through trade relations. Thus, wealth is estimated through the relative quantity of imports at a site, or concentrated in loci within a site. Imported pottery is most commonly used as an index of wealth, and is especially valued by archaeologists as it bears information than can indicate the approximate date and region of manufacture.

The presence of imported goods at Swahili stonetowns, and the relative increase of imports during town growth and development, has led to the tautological assumption that stonetowns represented concentrations of wealth, even in the absence of data from non-stonetown contexts. Thus, one of the primary interests in the investigation of earth-and-thatch settlements by the Pemba Archaeological Project was the illumination of patterns of wealth among these earthen architecture contexts within stonetowns and out in the countryside (LaViolette et al. 2003). Here I discuss the variety and relative quantity of imported goods excavated to date from town and village sites in Northern Pemba, relying on data from Fleisher's recent surveys (2001, 2003) and LaViolette, Fleisher, and Mapunda's 2002 and 2004 excavation season data, which has kindly been made available to me, as it is itself in the process of being analyzed and published.

Imported Pottery

Exotic ceramics were found at all four sites excavated by the Pemba Archaeological Project (LaViolette et al. 2003; LaViolette et al. 2004), and at almost all sites excavated by Fleisher during his 1999 survey season (Fleisher 2001, 2003). Pottery types and their associated time periods and origins are presented in Table 8.2.

The variety of imported ceramics found in Pemban sites is congruent with other ceramic assemblages found along the Swahili coast. The most common types prior to the twelfth century (Sasanian-Islamic – see Figure 8.3, White Glazed, Siraf, and some Late Sgraffiato) originated from the Port of Siraf in the Persian Gulf; later, imports include wares likely obtained from Red Sea ports (Fleisher 2003:274-275). Thus, expansion of the Indian Ocean trade network observable at other Swahili sites around A.D. 1300 also reached the shores of Pemba Island.

Wright (1993) uses an imported sherd ratio (the number of imported sherds per every 100 local sherds) as one indication of the relative amount of imports within a ceramic assemblage, which can be used as a

Table 8.2. Chronology of Imported Pottery Types found on Pemba. From Fleisher 2003:267-274, and Horton 1996:271-310.

Imported Ceramic Type	Attributes and Form	Date of Production	Source
Sasanian-Islamic	White fabric with blue to turquoise glaze; mainly large jars	8th-10th century	Persian Gulf
Olive Green Glazed	Hard grey fabric with olive glaze	8th-9th century	Far East
White Glazed	Yellow-white paste with white to cream glaze; small bowls	Mid-9th to 10th centuries	Persian Gulf
Siraf Storage Jars	Unglazed earthenware; large storage jars	9th-11th centuries	Persian Gulf
Late Sgraffiato	Variable: pink-orange paste with sand temper, white slip incised to create designs, topped with lead glaze; variable forms	11th-16th centuries	Persian Gulf
Black-on-Yellow	Yellow glaze on interior decorated with wavy black lines; bowls	13th-14th centuries	Southern Arabia
Celadon	Longquan greenware (grey fabric, green glaze); light brown glaze greenware	13th-16th centuries	Far East
Underglaze Painted Polychrome	Polychrome painting under clear glaze;	Early 14th century	Southern Arabia
Monochromes	Blue or green glazed, often with pock marks or bubbles;	Mid-14th century onward	SW Asia
Blue and White	Hard white fabric with blue and white decorations on exterior surface	14th -17th centuries	Far East

proxy for access to Indian Ocean trade. Use of common measures such as the imported sherd ratio and imported sherd density fosters cross-site comparisons that are useful at both the local (Pemban) and regional (Swahili coast) scale (see Table 8.3). For example, imported sherds appear to be much more common at the town of Tumbe compared to its neighboring village Kimimba before A.D. 1000, and Tumbe's imports are only slightly less common than those at Shanga, at that time a growing coastal trade center (Horton 1996). Some units within Tumbe yielded more imports and therefore higher (>5.0) imported sherd ratios, demonstrating probable inequality in access to Indian Ocean resources across the site.

Like the local wares, imported ceramics at this time are predominantly large jars. These were probably containers carrying liquids or semi-solids from the Near East, including date syrup and sesame oil (Horton and Middleton 2000:77). After their contents were consumed, these jars were likely used as food storage or preparation containers. I suggested in Chapter 5 that the predominance of pearl millet, along with sorghum and finger millet, before A.D. 1000 likely required more liquid preparations, including porridge (*ugali*) and beer (*pombe*). While the size and shape of local Tana pottery lends itself well to making beer and porridge, among other use, the use of exotic pottery in culinary activities cannot yet be ruled out.

By the founding of Chwaka, however, a different pattern emerges. First, Chwaka's imports are less common compared to Tumbe's, but sherd density is higher. Second, Chwaka at this time may be secondary to another Pemban trading center, Mduuni, based on the higher import ratio at the latter (Fleisher 2003:276). Third, the only village sampled during Chwaka's life-

Figure 8.3. Sassanian-Islamic pottery excavated from northern Pemba Island. Photo courtesy of the Pemba Archaeological Project.

time, Kaliwa, has an imported sherd ratio double that of the early village of Kimimba, and much closer in value to that of Chwaka. This differs considerably from the relationship observable between Tumbe and Kimimba, where Tumbe was clearly more integrated into the Indian Ocean trade network than Kimimba. In fact, Kaliwa's imported sherd value suggests that urbanization did not serve to differentiate Chwaka from the countryside in term of access to trade goods and wealth, but rather that urbanization may instead coincide with the increased availability of foreign goods to all Swahili communities, stonetowns and villages alike.

Table 8.3. Imported Sherd Ratios and Densities at sites on Pemba (in bold) compared with other contemporaneous coastal sites. From Fleisher 2003:276-277; LaViolette et al., 2003; Wright 1993.

Chronology	Site	Imported Sherd Ratio (# imports per 100 local sherds)[a]	Imported Sherd Density (sherds/m^3)
	Tumbe (town)	3.6	9.8
7th -10th centuries	**Kimimba** (village)	.2	.9
	Shanga (town)	4.2	5.3
	Chwaka (town)	.6	11.7
11th-13th centuries	**Mduuni** (town)	2.1	26.9
	Shanga (town)	6.5	-
	Kilwa (town)	2.0	-
14th-16th centuries	**Kaliwa** (village	.4	9.4
	Chwaka (town)	1.1	31.5

[a]Note that these figures come from different sized ceramic assemblages, thus some differences in value may be attributable to sample-size effects.

And fourth, ceramic forms after A.D. 1000. shift from mostly jars to bowl forms (Fleisher 2003, 2010). As I demonstrated in Chapter 6, the use of rice eclipses that of African grains during this time, and the use of bowls may be linked to the serving of rice. I will return to this use of ceramics in Chapter 9 while discussing Fleisher's (2003, 2010) hypothesis that food sharing and feasting may have been used by emergent Swahili elites to secure and maintain high standing within the community.

Taken together, the imported sherd data suggest that access to trade goods was concentrated within certain towns along the coast during the seventh to tenth centuries. This expanded, however, after A.D. 1100 to reach a larger proportion of the coastal population. If imported sherds signal access to the Indian Ocean trade network, and by association the accumulation of wealth, then the limited village data we have at present show that wealth was not concentrated in stonetowns during Swahili urbanization, but rather that wealth was available to rural as well as urban Swahili, although perhaps at a different scale. Thus, it may be that previous notions of rural Swahili as homogenously poor underestimated their connection with regional economies tied to Indian Ocean trade.

Other Imports: Glass and Beads

Several other important artifact classes are present on Pemba, as in major centers along the Swahili coast. Pieces of glass bottles, beakers, and flasks are among the more common of these imports. Surprisingly, the highest concentration of glass shards comes from the earlier sites of Tumbe and Kimimba, but glass was common at both town and village sites in all time periods (Fleisher 2003:314 Table 7.1; LaViolette et al. 2003). There appears to be little evidence to suggest that imported glass was restricted to use within towns, but rather that some rural households had access to glass imports as well.

Fleisher (2003:322) reports the presence of five rare imported beads at Kimimba, the early village with low imported pottery abundance. Two are early carnelian forms likely made in South Asia, and the remaining three are "drawn blue glass with white striped trails … [for which there] are no parallels … at other coastal sites" (Fleisher 2003:322). Glass beads are much more common along the coast, as demonstrated by the 213 glass beads at Tumbe (compared to 16 carnelian ones) and the 359 glass beads at Chwaka (compared to 17 carnelians) excavated in 2002 and 2004. While glass beads are more abundant in town contexts, the presence of both glass and carnelian beads in villages through time implies that village residents had access to these Indian Ocean trade items.

Summary

Evidence for Pemban access to Indian Ocean trade items comes in the form of imported pottery, glass, and beads, as well as other artifact classes for which the 2002 and 2004 analyses are not yet complete. Ceramics imported before 1000 A.D. appear to have been restricted to distribution within towns, as indicated by the disparity in imported sherd ratios between Tumbe (3.6) and Kimimba (.20). This pattern does not hold for glassware or beads, however, which represent significant finds at the village of Kimimba. Within the town of Tumbe, Fleisher and LaViolette (2013) note that ceramics were distributed quite evenly across earthen households, as were artifacts linked to production, which taken together suggest that Tumbe "flourished for centuries balanced between engagement in a domestic economy and a long-distance one" (ibid, 1166). The food economy from Tumbe seems to echo that interpretation, given the strong presence of pearl millet alongside the low incidence of Asian rice.

A.D. 1000 imported pottery appears much more evenly distributed among settlement types; although urban contexts contain relatively more foreign ware, there is notably less difference between the imported sherd ratios of Chwaka (1.2) and Kaliwa (.43) during the fourteenth to sixteenth centuries than was observed between the first millennium sites. Glass and beads were also present in both urban and rural contexts during urbanization. Taken together, the import data draw a picture of more evenly distributed wealth among different settlement sizes than predicted by current models of Swahili urbanization, which focus on the stonetown as the locus of control over Indian Ocean trade and the wealth it offered (Horton and Middleton 2000; Kusimba 1999; Wright 1993).

Having surveyed the local production and foreign consumption patterns among sites on Pemba Island, I now turn to integrating the food production data, presented in Chapters 5-7, into the regional economy in northern Pemba.

CHAPTER 9: CONCLUSIONS

Introduction

In this final chapter, I summarize the major findings of this study and place the food production system in northern Pemba, as shown through archaeobotanical analysis, in context with local and regional patterns of urbanization. I end the chapter, and the book, by discussing the limitations of this study and suggesting directions for future research.

Archaeobotanical Evidence of Swahili Diet and Plant Food Economies

This study uncovered a wealth of data regarding early Swahili diet and plant food production, and reports one of the richest archaeobotanical assemblages known from eastern Africa. Here I review the data reported in previous chapters, beginning with the natural and cultural factors that may have contributed to the preservation of plant remains in Swahili contexts.

Preservation

The most basic finding of this archaeobotanical study is also possibly the most important: these tropical Swahili archaeological sites all contained an abundance of identifiable charred plant remains, including food plants. This requires explanation considering that plant remains are widely considered to be rare in tropical archaeological contexts (Hather 1994) and that previous attempts to recover identifiable plant materials from eastern Africa have largely met with failure (Ambrose 1984; Reid and Young 2000; Robertshaw and Wetterstrom 1989; Wetterstrom 1991; Young and Thompson 1999; but see Wetterstrom and Wright 2007 and Wright 1984, 1993 for success stories on the Comoros Islands). The remarkable preservation in Pemban contexts is likely attributable to the regular burning of fires close to domestic contexts, and the deposition and rapid burial of household debris in midden contexts within these relatively dense settlements. Chwaka midden deposits accumulated rapidly and show little signs of disturbance from overlying deposits, unlike those at Tumbe, and this rapid deposition, coupled with dense occupation, may account for the outstanding level of preservation among Chwaka samples (Walshaw 2005). Additionally, sedimentological differences among sites may account for differences in the quality of preservation noticeable in the archaeobotanical, and particularly the faunal, remains (Cain 2005; LaViolette et al. 2003, 2004).

The issue of how plant remains become carbonized and preserved archaeologically (botanical taphonomy) has received close attention (Fuller et al. 2005; Miksicek 1987; Miller and Smart 1984) and warrants discussion here. The high ubiquity of wood charcoal suggests that fire residues were commonly dumped in middens, along with any particles dumped into the fire from cooking and cleaning activities. Moreover, the high frequency of fruit and nut remains implies that these taxa, particularly coconut, were not only consumed, but may have served as an important fuel source. Coconut likely had many household uses, including as a food source, particularly in the form of fresh water and dried storable coconut flesh (copra). It is also a source of fiber for cordage, raw material for scoops, bowls, and other tools, and finally as fuel (Ohler 1999b). Moreover, coconut trees are long-term producers, and stands may be passed down between generations, or sold in times of need; in this way, coconut palms may serve as a form of wealth.

Seeds and processing waste found in household contexts are considered byproducts of post-storage processing and meal preparation, where waste could be dumped intentionally into a fire or midden, and grains more likely entered by accident. Special areas of abundance, such as the pearl millet, rice, and cotton concentrations discussed in Chapters 5 through 7 may be the result of an accidental burning of foods stored in the house. The preservation of some plant remains outside of midden contexts, such as on house floors, could be due to accidental fire, wall fall, or the deposition of burned materials close to cooking hearths or *mofa*, clay pot ovens. Finally, the few small wild and weedy seeds observed in the Pemban samples may have entered the archaeobotanical record as primary food products or as by-products of crop processing waste or gardening waste.

Subsistence Patterns Across Time and Space

Several time-related patterns emerge from an investigation of the plant food remains. In general, the archaeobotanical assemblages of the first millennium sites of Tumbe and Kimimba more closely resemble each other, and to a lesser extent Kaliwa, than either resembles Chwaka, the eleventh to sixteenth century stonetown. This is best exemplified by the frequency of rice and fruit taxa (see Figure 9.1). Period 1 deposits at Tumbe and Kimimba contained few rice grains and chaff. Rice is present at only 30% of samples from Tumbe, and is absent from Kimimba's samples. This early Tumbe rice may represent contamination from upper Mazrui period levels, although this is unlikely. Rice is 100% ubiquitous, however, at both Chwaka and Kaliwa, with rachillae being slightly more common than grains among non-storage contexts at all sites where rice is found. Thus, while rice apparently was more accessible to the residents of Tumbe compared than those at Kimimba, patterns of rice distribution for the second millennium sites are best explained as an increase in access or popularity through time, as opposed to differences in the food production activities at each site.

Coconut is sparsest at Kimimba, while all remaining sites show a high coconut ubiquity. At Kimimba, both coconut and rice are significantly under-represented compared with the contemporaneous town of Tumbe, suggesting that Tumbe residents had greater access to these introduced commodities than people at Kimimba. Chwaka, by comparison, is rich in coconut as well as other fruit taxa. Non-coconut fruit and nut fragments

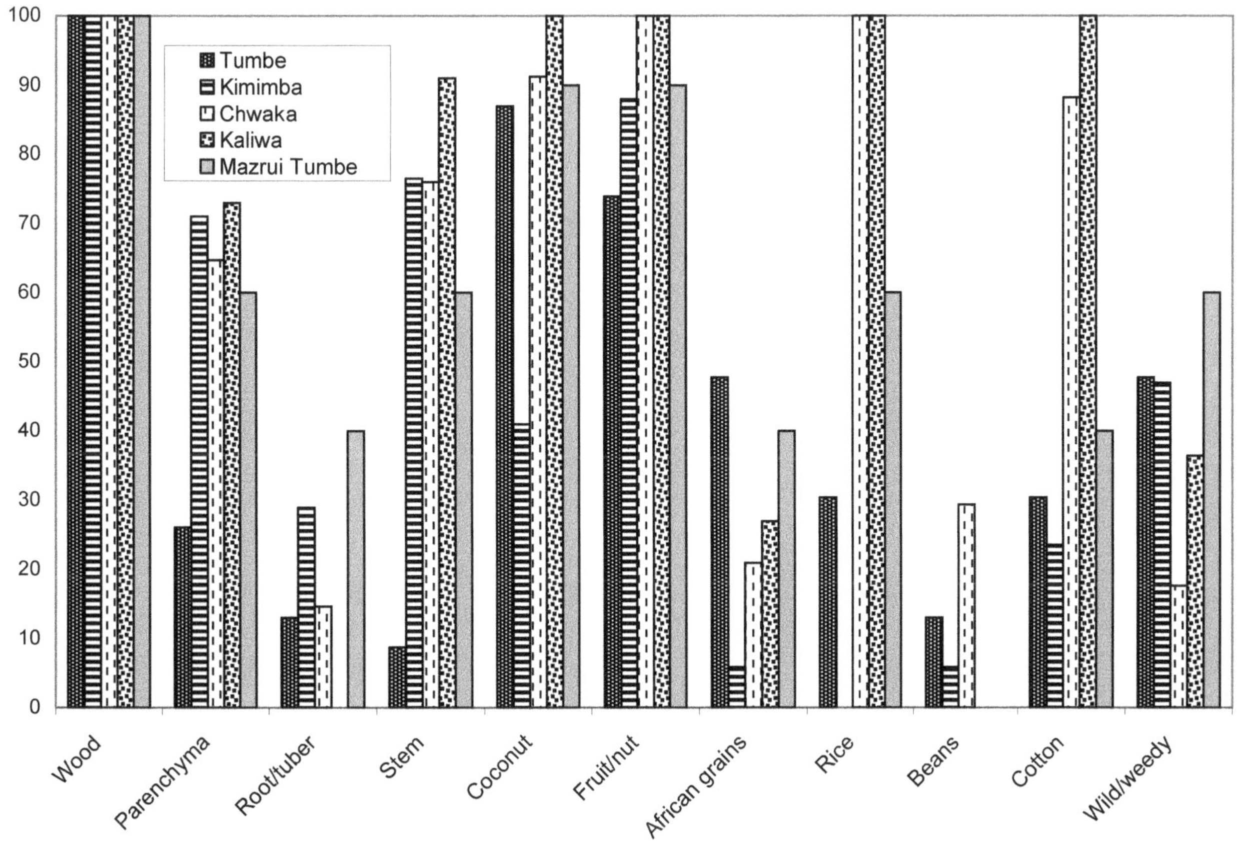

Figure 9.1. Ubiquity measures for major botanical categories across all sites.

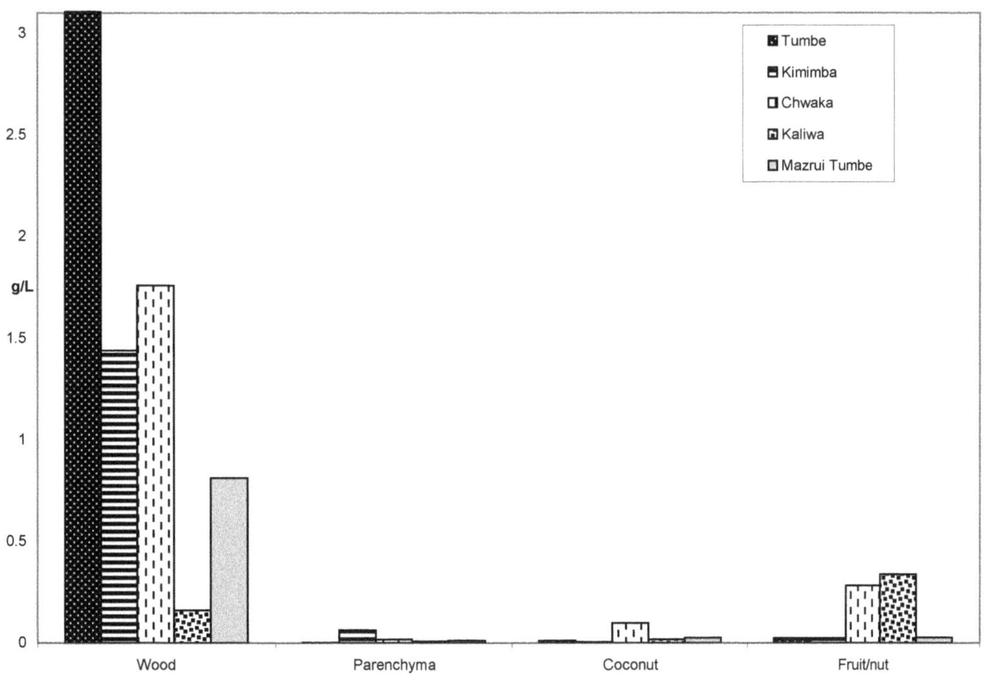

Figure 9.2. Density of non-seed macrobotanicals across all sites.

are found in low density in Tumbe (all time periods – see Figure 9.2) and Kimimba, but constitute a significant proportion of the archaeobotanical assemblages of Chwaka and Kaliwa, with the latter producing apparently a greater variety of taxa than observed for Chwaka. Root/tuber fragments were present at both Tumbe and Kimimba, but sparse at Chwaka and absent from Kaliwa. While cotton is present at all sites in all time periods, the ubiquity of this taxon varies from 30% and below at Tumbe and Kimimba to virtually 100% at Chwaka and Kaliwa, with a significant concentration in an early Chwaka deposit. This pattern of cotton distribution closely follows that observed for rice.

By the founding of Chwaka in the eleventh century, however, cotton production had intensified, as had virtually all other arenas of plant production, most notably rice and coconut. The appearance of a cotton seed concentration (some bearing lint) at the earliest levels of Chwaka bears testimony to the importance of cotton in the economy at the founding of the town. This is interesting considering the assertion by Horton and Middleton (2000:13) that cotton was imported by the Swahili from Indian sources; according to the archaeobotanical record, local cotton production clearly existed alongside importation of foreign cloth. The co-occurrence of local production and importation of finished products also exists for iron-working, so may not be unexpected among the Swahili. Other intensified products at Chwaka include coconut, other fruit and nut taxa, and rice. The increase in charred coconut at Chwaka may be due to the use of coconut shell as fuel, and may signal the increased use of other coconut products as well, for food, fiber, and tools. At this time, rice dominates the grain element of the archaeobotanical record, indicating an intensification of rice production by the eleventh century. The abundance of both rice grains and rachillae (spikelet bases) supports the production of rice plants, and not merely the consumption of imported rice. Pearl millet and other African grains all but disappeared from the economy of Chwaka until the latest century in the sequence (Walshaw 2010), when Kaliwa also demonstrates millet use. A return to these drought-tolerant crops may signal a change in the timing and duration of the annual rains, or indicate that landscape depressions suitable for wet rice cultivation were limited due to environmental conditions, previous efforts toward extensification, or possibly political control over the countryside.

There are spatial patterns obvious as well among the Pemban archaeobotanical assemblages. Pearl millet appears to be most common at Tumbe, in both Periods 1 and 5, signaling a possibly spatially-dependent trend through time. Whether this is due to mixing between levels can only be ascertained by directly dating the materials from the Mazrui Tumbe layers, as pearl millet from an early stratum at Tumbe was directly dated to the seventh century. It remains possible that preservation conditions in the microenvironment of Tumbe are more amenable to grain preservation, but this is unlikely considering Tumbe's close proximity to Chwaka, where African millets were found only towards the end of the sequence despite exceptional preservation of other plant remains.

In sum, the early Pemban economies (prior to A.D. 1000) seem focused on pearl millet, with contributions from legumes, coconut, and other fruit and nut taxa. Rice was introduced in this time period, but occurs infrequently (Walshaw 2010). The use of coconuts at this time may be attributed to either the introduction of Malaysian coconuts or the intensification of wild stands of sea-dispersed coconuts. Cotton is quite sparse throughout samples from this time period, but it was present at both Tumbe and Kimimba, likely for use in fiber or oil production. Significantly no ceramic or stone spindle whorls are found dating to this period, on Pemba or elsewhere along the coast (e.g., Shanga; Horton 1996), which raises the possibility that organic spindle whorls were employed, a subject I addressed in Chapter 8 alongside the discussion of Pemban craft production.

Plant Food Production Relations between Town and Country

Chapter Seven was dedicated to a discussion of the economic relations between the towns (Tumbe, Chwaka) and contemporaneous neighboring villages (Kimimba and Kaliwa, respectively) to determine what, if any, changes in plant food production accompanied urbanization. I briefly summarize the results here. While the transition from an African grain economy to one centered upon Asian rice cultivation signals intensification and specialization, there is little evidence to support the hypothesis that village farming communities were providing the larger towns with grains for food. This does not preclude the possibility that grain was collected in small amounts as tax or tribute, but the archaeobotanical record does not indicate a producer-consumer relationship between the villages and towns represented by this study. Moreover, it does not appear likely that towns possessed administrative control over rural food production, or that certain individuals collected grain for redistribution within the town or elsewhere. Thus, rice is accepted in the Swahili economy without a significant alteration to agricultural production strategies. In a sense, while rice is botanically Asian, its use among the Swahili is undoubtedly African.

Further archaeobotanical work is required at other Swahili settlements of all kinds to determine if this pattern persists across the Swahili coast. It would be particularly interesting to determine how the largest and most complex stonetowns (e.g., Kilwa, Shanga, Manda) met their plant food requirements. Also, an examination of food production practices at Mombasa would be instructive because of the historically documented food production relationship between Pemba and Mombasa. The archaeobotanical evidence to date does not imply that any of the four sites on

Pemba were net producers of grain destined for consumption at Mombasa. Having summarized the major findings of this study, and put forward a model of how ancient agriculture may have operated among these Swahili communities, I now turn to applying this data to an understanding of emergent urbanization on northern Pemba.

Prestige Economies and Food Production: the Case of Swahili Urbanization

Founding Settlements and Economies: Period 1 (A.D. 700-1000)

Rice and radiocarbon dated pearl millet from seventh century Tumbe indicate that agriculturalists were among the first settlers on Pemba. But the relative rarity of grains at the contemporaneous village site of Kimimba, coupled with the importance of root/tuber and fruit/nut taxa, imply different subsistence strategies at work there. While preservation conditions at Kimimba were not ideal, indicated by the rarity of animal remains (Cain 2005), the recovery of intact tubers, seeds, and other plant tissues suggest that differential preservation alone does not account for the archaeobotanical disparities observed between Tumbe and Kimimba. Not only does this pattern imply economic independence between the two sites, but, taken together with local ceramic and settlement data, it may indicate differences rooted in ethnicity.

As discussed above, much variability can be observed among decorative elements within Period 1 local (Tana) ceramics from northern Pemba (Fleisher 2003:238-245). It is possible this is the result of multiple migrations to the island from different communities along the coast, some of whom may not have been Swahili (Fleisher 2003:403). According to linguistic analyses the Swahili language was developing during the sixth to ninth centuries A.D., among communities along the coast of present day Kenya (Nurse and Spear 1985). While it is clear from the architectural and material culture evidence that Kimimba maintained many affinities with nearby Swahili settlements (Fleisher 2003; LaViolette et al. 2003), notable differences in subsistence strategy suggest that Kimimbans relied more heavily on fish and local wild plant resources than their millet-cultivating, cattle-herding neighbors at Tumbe. This is one potential indicator of an ethnically heterogenous landscape existing on Pemba during its initial colonization.

Settlement pattern data, yielded through survey and recent excavations, also illuminate our understanding of the first settlements on Pemba. Fleisher (2001, 2003) discovered many communities of differing sizes existing in Period 1 during his survey of northern Pemba, a pattern not predicted by earlier studies relying on the few large coastal settlements known from this time (e.g., Horton 1996, Kusimba 1999). While the two towns recorded for northern Pemba during this time (Tumbe, Bandarikuu) fit well with other known coastal settlements, most of the communities discovered during survey were in fact small hamlets and villages, dotted throughout the landscape and located up to 6 km away from the coast (e.g., village of Gombe) (Fleisher 2003:Chapter 5). Fleisher's rank-size distribution analysis of these settlements indicates the absence of site hierarchies among the initial settlements, and little integration among the economic and political realms of each community (Fleisher 2003:393-396). The food production data summarized at the beginning of this chapter resonate well with this image of settlement autonomy during Period 1, in which Tumbe's relative wealth did not trickle down through rural channels, in spite of the regional communication networks that must have been operating in this densely settled landscape with shared ceramic traditions.

Another way that food production intersected with local economies on Pemba at this time was the potential use of beer, made from sorghum, finger millet, and/or pearl millet, in feasting. Feasts, communal and ritual food-sharing events, are commonly acknowledged as an effective way to entice and reward acts of labor in agrarian societies (Dietler and Herbich 2001). As described in Chapter 7, seasonal labor would have been crucial to the success of major crop harvests, including African grains and cotton, foundations of the subsistence and craft economies on Pemba. The importance of African grains during Period 1, particularly pearl millet, coupled with the predominance of locally produced pots and bowls in the ceramic assemblage (Fleisher 2010), is consistent with the preparation and sharing of beer. Individuals that could command more agricultural and craft production labor would have increased access to Indian Ocean trade goods, reinforcing the value of the feast and generating the inequality observed among wealth distribution patterns on Pemba during Period 1 (Fleisher 2003:417-419).

With an increasing commitment to Islam, however, the use of beer in feasts may have waned due to religious alcohol prohibition, although adherence to this facet of Muslim law has been differentially applied throughout the Islamic world (Heath 2000:90-98). This may, in part, explain the rapid replacement of fermentable millets with Asian rice, which, as I discuss below, was itself well suited to the physical and political environments of northern Pemba during Periods 2 and 3 of Swahili inhabitance.

Emerging Elites and Competitive Feasting: Periods 2 (A.D. 1100-1300) and 3 (A.D. 1300-1500.)

Similar to Period 1, the archaeobotanical data recovered from second millennium Swahili contexts in northern Pemba articulate well with local production, consumption, and settlement data, and serve to enrich our understanding of urban forms emerging by A.D. 1100. At the founding of Chwaka by A.D. 1050 the archaeobotanical record clearly indicates reliance upon rice, coconut, and cotton; virtually absent are the African millets so important to Period 1 Tumbe. While Chwaka was founded close to the recently abandoned

Tumbe, these two towns share very little in the way of material culture and food production, suggesting that Chwaka was founded by a new population within a century or less of Tumbe's abandonment (LaViolette et al. 2003; Fleisher and LaViolette 2013).

Settlement pattern data for Period 2 generated by Fleisher's survey reveal the emergence of new villages that rapidly grew to be stonetowns, concomitant with a decrease in rural population (Fleisher 2003:406-407). This differs remarkably from the variety of dispersed communities that characterized Period 1. These developing towns apparently grew at the expense of the countryside, a phenomenon that raises several interesting questions about how towns attracted new residents, but answers others as well. For example, the migration of rural households into towns may help explain the retention of a household-based plant food production system in northern Pemba's emerging towns. If towns were populated with food producers, and few remained in the countryside, then it is reasonable to conclude that towns were capable of producing their own food, with the majority of residents probably working in nearby fields. Moreover, as population pressure increased, people may have moved into more densely packed towns to leave valuable agricultural land for farming, as opposed to habitation, a practice that is observable today on Pemba's Micheweni Peninsula. Here, coral outcroppings and thin, reddish, soils make for marginal sorghum farming at best, but workers living in this region cross Micheweni Bay daily to work in the rich rice fields near to the standing ruins at Chwaka (H.M.Hajj, personal communication 2004). Thus, not only might the growing population of Chwaka have been stocked with farming households, but this migration pattern implies that the countryside was becoming relatively impoverished in food producers. This depopulation of the countryside over time increased as new stonetowns arose after A.D. 1300 (Fleisher 2003:407), potentially making the food production of towns even more independent as a result.

But what about urban life attracted the countryside population? Fleisher (2003: 417-419) argues that competition arose among emerging stonetowns and their elites for rural migrants and their craft production labor; popular support was an essential part of gaining and maintaining elite status. Competing elites vied for immigrants through acts of public generosity, predominantly feasting (Fleisher 2003: 417-422; Fleisher 2010), and used Islam as a means of maintaining public support in the face of emerging social inequality (Wright 1993). Thus, villagers may have been drawn to centers based on the generosity of their leaders, and could have moved to new towns if dissatisfied, perpetuating the need for elite generosity through feasts that may have resembled what Dietler (2001:82-85) describes as "patron-role" feasts. Feasting has recently gained remarkable attention in the archaeological literature as a political tool for gaining power (e.g., Dietler 1996) and an adaptive force in human prehistory (e.g., Hayden 2001), and case studies from around the globe are well represented in various themed volumes (Dietler and Hayden 2001; van der Veen 2003b; Wiessner and Schiefenhövel 1996). While interpretations of feasting often presuppose the presence of exotic, high-valued, or domesticated goods, in some societies feasts incorporated wild plants and animals that likely possessed important symbolic value. For example Sub-Mound 51 at Cahokia was rich in deer, swan, and local crop and fruit plants, yet contained little maize, an introduced and important food, in abundance (Kelly 2001; Pauketat et al. 2002). Yet another variation on feasting is that of amount as opposed to kind: the ability to host a feast in which guests ate in abundance seems to have been an important concept among some African societies (Goody 1982). As I presented in Chapter 3, early visitors to the eastern African coast reported eating rice with meat sauce as well as tropical fruits during meals with important local elites, suggesting that everyday food bore importance in special, and potentially ritual, circumstances.

Three lines of evidence can be shown to support the argument that competition among Swahili patrons encouraged acts of generosity that may have taken the form of feasting. Competition between towns is evident in their regular spacing throughout the countryside, which Fleisher (2003:411-412) measured using nearest-neighbor analysis. Fleisher's (2003:413) data suggest that:

> settlements on Pemba during Period 3 are ... considerably well-spaced; that is, second-order towns were not clustered around higher order stonetowns, but regularly spaced throughout the island, extending into the less fertile ... southern half of the island. These data suggest that new settlements on Pemba were always founded with an eye towards others-providing a measure of the increasingly competitive environment on the island.

Additional evidence comes in the form of ceramics. As was noted previously in this chapter, both local and imported ceramic assemblages show a shift in form, from predominantly pots and jars to largely bowls and platters. Local ceramics in particular bear increased ornamentation on the inner surface, including painting and burnishing. This increase in decorative serving ware is consistent with the presentation of meals in display contexts, including feasts (Fleisher 2010). Moreover, open vessel imports indicate the local demand for these items as serving or display dishes, because they could not be used to carry oils or syrup as the Period 1 jars were. Also, the display of exotic wares in the wall niches (*zidaka*) of mosques and tombs may have served as public and ritual reminders of the generous acts of elites (Fleisher 2003:419).

Lastly, the archaeobotanical evidence that rice dominates the Swahili diet during Periods 2 and 3 accords well with a hypothesis of feasting, but in amount rather than in kind (Goody 1982:78). That is, all households consumed some rice, and knew its value, so would be duly impressed if heaps of rice were to be served by local elites. There is as yet no archaeobotanical evidence to suggest differential access to exotic or rare foods among certain residences of Chwaka, which is notable considering that feasts displaying access to luxury goods are known elsewhere in the Islamic world (MacLean and Insoll 2003). That the decline of Swahili stonetowns on Pemba coincides with an increase in millet consumption is perhaps telling of the central role that rice feasting may have played in local politics.

Urban Food Production in Sub-Saharan Africa

Having described the Swahili food production system and placed it in perspective with other elements of urbanization in northern Pemba, I will now return to some of the main theoretical issues surrounding urban food production that were raised in Chapter 2. The archaeobotanical evidence to date indicates the presence of a household-based food production system, in which most households likely produced their own food unfettered by urban administrative control. While town and country grew largely the same crops after A.D. 1100, it does not appear that rural populations provisioned urban ones. The few, if any, urban elites or religious leaders who did not produce their own food might have bartered for or received grain from community members as part of a ritual relationship, probably cemented with feasting.

The local agricultural economy that supported the growth of Chwaka was in place literally from the inception of this settlement as a village by A.D. 1050, with few observable changes through time until a slight shift towards African millets becomes apparent by fifteenth century. This relative stasis is significant because urban systems are expected to alter food production practices to suit the needs of an emerging city or state. This may take the form of standardization of production (to facilitate redistribution), centralization of processing (to bring labor and product under administrative control), and/or taxonomic specialization (to meet specific trade or provisioning requirements) during the growth of the city and its leadership (Zeder 1991, 1999). Conversely, it appears that in the single Swahili stonetown for which we have adequate data, the urban structure is pre-adapted to the form and substance of agricultural production. Thus, the stonetown grew around the agricultural system instead of altering it, leading to extensification and, likely, intensification (through irrigation) of production without substantially changing its form. On Pemba, this retention of agricultural traditions may have been possible with the manageable size of Chwaka relative to the fertile countryside; in other words, Chwaka may have never outgrown its food supply. Different patterns, and ones that more closely resemble the production of food in Mesopotamia and Mesoamerica, might be expected at larger stonetowns along the coast, such as Kilwa, Manga, and Shanga, among others.

Yet the agricultural patterns noted for Pemba have some resonance with other examples of African urban food production. It is reminiscent of that described in Chapter 2 at Jenné-jeno, where fish, African rice, and local wild grasses dominate the subsistence all throughout the occupation sequence, and where the growth of the population is attributed to extensification of food production and redistribution through Jenné-jeno's market (McIntosh 1995b). Similarly, Aksum's food economy was largely market-driven as indicated by preliminary analyses by Boardman (1999, 2000), and it appears likely that urbanites obtained food from farmers and gardeners who brought their wares to Aksum's market. The examples available from Zimbabwe tradition sites indicate a specialization in cattle production during the rise of the state, and the introduction of domestic grains into a wide plant food subsistence base, but mechanisms of redistribution as are yet unknown (Jonsson 1998).

Thus, among the sub-Saharan examples presented in this study, including the case of the Swahili, a pattern begins to emerge. First, urban and state growth is aided by the prior existence of an adequate food production system, which is not greatly altered in the process of urbanization. Second, urban sites facilitate food redistribution through market and/or ritual/religious relations, and no centralized framework exists for the provisioning of non-food-producing sites or individuals. Third, food production is increased through extensification, or the expansion of agricultural land. Agricultural intensification is rare among indigenous forms of sub-Saharan agriculture, possibly because they may leave behind little or no archaeological evidence (Soper 1997), but terracing has been reported after A.D. 1500 at the Nyanga complex in eastern Zimbabwe (Summers 1958) and at Engaruka in northwestern Tanzania (Sutton 1978, 1986).

In the case of the Swahili, Asian rice cultivation likely occurred in natural, possibly leveled, landform depressions where rainwater collected to form a moist paddy. Today, Swahili farmers dig small canals around each rice field to maintain moisture levels. This practice may constitute a form of agricultural intensification, where land productivity is increased through alteration of the growing environment, i.e., terracing, raised fields, and irrigation (Boserup 1965; Kirch 1994). Until more is known about ancient Swahili agricultural techniques, this postulated means of intensification remains to be verified. Yet the growth of towns like Chwaka undoubtedly required extensification of land use in addition to any intensive rice agriculture, because the hilly nature of Pemba presents as many ridges as depressions, and suitable rice farming land is not concentrated within a certain area.

These four varied case studies alone do not permit the recognition of an "African" mode of urban food production, and to do so might overly generalize, or wrongly characterize, a trend that is still emerging. But scholars should recognize that just as African cities and states are gaining interest as alternatives to a previously assumed cross-cultural urban phenomenon (LaViolette and Fleisher 2005; McIntosh 1999; McIntosh and McIntosh 1993), so should their food production patterns. Conversely, just as early African food production systems are becoming known for their novel pathways and timing (Marshall 1998; Marshall and Hildebrand 2002; Neumann 2003, 2005), so archaeobotanists and zooarchaeologists need to pursue these trends through time to the development of complex societies in Africa. African scholars are only beginning to reveal the various forms of complexity that took place on this continent, and can contribute significantly to understanding complexity at the global scale (McIntosh 1999). But first, archaeologists and archaeobotanists need to build comprehensive and comparable data sets by investigating local patterns and regional networks, one site at a time.

Limitations of This Study and Suggestions for Future Research

The primary goal of this study was to illuminate Swahili food production practices and their intersection with other production and consumption economies during the different phases of urbanization within one region. There are several areas in which this study fell short of its goal, largely with respect to sampling and archaeobotanical identification.

Sampling

The present sample set was necessarily limited to houses and middens located within identifiable settlements. Yet the early stages of crop processing, and certainly field management practices, would have taken place outside of settlement areas, near to or in the cultivation field. Only garden plots within settlements would have been close enough to contribute waste from all processing stages to household midden deposits. Thus, it is entirely possible that weeds, culm nodes, and other botanical waste generated by the early stages of crop processing were deposited in the field, where their likelihood of preservation and subsequent archaeological recovery is very low.

Today, Swahili farmers conduct many of these early-stage crop processing activities in field houses, which are semi-permanent structures erected near fields that are used during critical points in the agricultural year. The role of field houses in ancient Swahili plant food production is virtually unknown because archaeologists have difficulty recognizing such ephemeral, isolated structures (but see Fleisher 2003:134). Thus, documenting the early stages of crop production archaeologically is contingent upon identifying field houses (Posnansky 1984). While Fleisher recognized four field houses in northern Pemba during his survey on the basis of site size (<.10 hectares) and a narrow range of local pottery (2003:134, 137), archaeobotanical signatures of events taking place in field houses may prove valuable to aiding in the recognition and identification of these structures in the archaeological record. I have undertaken ethnoarchaeological investigations of a recently abandoned field house to address these questions.

Identification

This study was limited by our present lack of knowledge concerning the botanical identification of some local plants, particularly fruit and nut taxa. This is not surprising considering that Swahili archaeobotany is emerging as a new field of interest, and that botanical collections in local herbaria are restricted at present. I hope that the findings reported in this study, particularly the abundance of macrobotanical remains, inspire further work. I advise that future archaeobotanical investigations are conducted in parallel with the collection of botanical comparative specimens and ethnobotanical knowledge. These collections should be duplicated in African and European or North American herbaria to facilitate identifications while in the field as well as in the home laboratory. Ultimately, scholars should work together to create adequate facilities for the analysis and identification of archaeobotanical materials at eastern African institutions themselves, for example at the University of Dar es Salaam, or the National Museums of Kenya.

Regional Specificity

The archaeological sites considered in this study were limited to two towns and their neighboring two villages on Pemba Island, as an example of how town and country interacted during Swahili urbanization. Thus, the data and the interpretations deduced from them are most directly applicable to the urban question as it pertains to northern Pemba. The patterns observed here may not hold at different sites along the coast, where individual stonetowns underwent different cultural and economic histories. Pathways to urban food production among Swahili towns may have followed different trajectories: possibly with respect to the plants used, the chronological or spatial distribution of botanical patterning, or the relationships between stonetowns and their hinterlands.

This is an exciting period of research in eastern Africa, with fundamental assumptions being questioned about the nature of coastal communities and their interactions with the Indian Ocean world (Fleisher et al. 2015; Helm et al. 2012). Archaeobotany has a prominent role in these debates as well as those related to the early movement of African and Asian plants across the Indian Ocean (Fuller et al. 2011). While this first study of Swahili plant subsistence and food production was necessarily limited in scope, recent work has expanded our knowledge of plant use along the eastern African coast, including the drier regions in the south (Pawlowicz 2011). This is bolstered by the incorporation of multiple lines of archaeobotanical research at the 14th-15th C site of Songo Mnara in the Kilwa Archipelago. Here, phytolith analysis (Sulas and Madella 2012; current analyses by Hayley McParland) has complemented macrobotanical investigation (Walshaw 2015; McParland and Walshaw 2015) of the use of space in both private (domestic) and public (open-air; ritual architecture) places, and in the wider environs (Stoetzel 2014). As more evidence is brought to light, we may be able to speak of regional plant-use patterning along the Swahili coast, and perhaps more data to explore the observation that the later stages of occupation at Chwaka saw a resurgence in the importance of African grains.

Epilogue: The Future of Archaeobotany on the Swahili Coast

In the decade since the inception of this study, numerous advances have been made in reconstructing ancient environments and lifeways on the Swahili Coast using archaeobotanical methods. Archaeologists are now more routinely adding botanical recovery and analysis to their large projects, including flotation for macrobotanical analysis and sediment sampling for phytolith analysis. Microbotanical methods are becoming powerful tools to address Swahili interactions with their natural and built environments. Phytolith analyses have been implemented with great success (e.g., McParland and Walshaw 2015; Pawlowicz et al. 2014; Stoetzel 2014; Sulas and Madella 2012.) Environmental archaeology, including botanical analysis, has challenged previous assumptions in the historical literature concerning past environments (e.g., Pawlowicz et al. 2014; Stoetzel 2014).

As archaeobotanical methods become routine in eastern Africa, it is important to build local resources to support both international and local projects. Currently, researchers are permitted to take flotation samples (light and heavy fractions) out of the country; in fact, lack of research facilities in Tanzania required me to do so for the study presented herein. However, for most artifact classes, Tanzanian Antiquities laws require special permission for removal of specimens for study abroad; the vast majority of finds remain in the country. Building resources for archaeobotanical research in eastern Africa will contribute to enlarging comparative collections and will work towards keeping archaeobotanical remains in the country. Together with Dr Thomas Biginagwa, I have helped establish an archaeobotany laboratory and comparative seed collection at the University of Dar es Salaam. We held an archaeobotany workshop in February of 2014 to train Tanzanian students in archaeobotanical identification and interpretation, in particular to support two MA students, Charisy Kawiche and Cecylia Mgombele , in their archaeobotanical research on agriculture along caravan trade routes in 19th C Tanzania.

The future of the field is bright. Archaeologists are increasingly adding macrobotanical and microbotanical analyses to their research programs. The range of research questions is expanding and specialists are applying state-of-the-art techniques. This is a very productive era in Swahili archaeobotany, and it will be exciting to see what the future holds.

Appendix A.
Major plant taxa referred to in the text, listed alphabetically by plant family. From Dharani (2002), Koenders (1992) and Williams (1949).

Family	Common name	Latin name and authority	Swahili name
Agavaceae	Sisal	*Agave sisalane* Perrine	Mkonge
Amaranthaceae	Amaranth	*Amaranthus* sp.	Mchicha
Anacardiaceae	Mango	*Mangifera indica*	Mwembe
Apocynaceae	Bungo	*Saba florida* (Benth.) Bullock	Mbungo
Araceae	Tannia, cocoyam	*Xanthosoma sagittifolium* (L.) Schott	Majimbi, Malanga,
	Taro	*Colocasia esculenta* (L.) Schott	Mayugwa
Arecaceae	Betel palm	*Areca catechu* L.	Mpopoo
	Borassus palm	*Borassus aethiopum* C. Martius	Mvumo/Mpapindi
	Coconut	*Cocos nucifera* L.	Mnazi
	Oil palm	*Elaeis guineensis* Jacq.	Mchikichi, Mjenga
	Date palm	*Phoenix dactylifera* L.	Mtende
Boraginaceae	East African Cordia	*Cordia africana* Lam.	
Brassicaceae	Mustard	*Brassica* sp.	
Bromeliaceae	Pineapple	*Ananas comosus* (L.) Merr.	Mnanasi
Chenopodiaceae	Goosefoot	*Chenopodium* sp.	Mpurele
Combretaceae		*Lumnitzera racemosa*	Kikandaa, Kkandaa-mwitu
Compositae	Noog	*Guizotia abyssinica* (L. f.) Cass.	
Convolvulaceae	sweet potato	*Ipomoea batatas* (L.) Lam.	Kiazi cha tamu
Dioscoraceae	yam	*Dioscorea* L.	Kiazi-kikuu, Mtubwi
Euphorbiaceae	Physic nut	*Jatropha curcas* L.	Mbono-mpaka,
	Cassava	*Manihot esculenta* Crantz	Mbono-pwani
		(syn. *Manihot utilissima* Pohl)	Muhogo
	Emblic	*Phyllanthus emblemica* L.	
	Castor bean	*Ricinus communis* L.	Mbono, Mbarika
Fabaceae		*Brachystegia* sp.	Myombo
	Hyacinth bean	*Lablab purpureus* (L.) Smartt	Mfiwi
	Horse gram	*Macrotyloma uniflorum* (Lam.) Verdc.	
	Tamarind	*Tamarindus indica* L.	Mkwaju, msisi
	Pea	*Pisum sativum* L.	Njegere
	Ethiopian pea	*Pisum abyssinicum* A. Braun	
	Cowpea	*Vigna unguiculata* (L.) Walp.	Kunde
	Mung bean	*Vigna radiata* (L.) R. Wilczek	Mchekcheke
Malvaceae	Cotton	*Gossypium* sp. (cultivated)	Mpamba
		Gossypium kirkii Mast. (wild)	Mpamba-mwitu
	Baobab	*Adansonia digitata* L.	Mbuyu, Mkuu
	Kapok	*Ceiba pentadra* L. Gaertn.	napingwa
			Msufi
Meliaceae		*Xylocarpus granatum* Koenig	Mkomafi, mtonga
Moraceae	Breadfruit	*Artocarpus altilis* (Parkinson) Fosberg	Mshelisheli
	Jackfruit	*Artocarpus heterophyllus* Lam.	Mfenesi
	Fig	*Ficus carica* L.	Mtini
Musaceae	Enset	*Ensete ventricosum* (Welw.) Cheeseman	Mgomba-tumbili
	Sweet banana	*Musa acuminata* Colla	Ndizi
	Plantain	*Musa paradisiaca* (L. pro. sp.)	Mgomba

Family	Common name	Latin name and authority	Swahili name
Myrtaceae	Clove	*Syzygium aromaticum* (L.) Merr. & Perry	Mkarafuu
	Rose apple	*Syzygium jambos* (L.) Alston	Mtofaa
Papaveraceae	Opium poppy	*Papaver somniferum* L.	
Pedaliaceae	Sesame	*Sesamum indicum* L.	Ufuta, Mfuta
Piperaceae	Black pepper	*Piper nigrum* L.	Pilipili-manga
Poaceae	Job's tears	*Coix lacryma-jobi* L.	Mtasubihi
	Indian Crabgrass	*Digitaria longiflora, D. Milanjiana*	Mkuse
	Finger millet	*Eleusine coracana* subsp. *coracana* (L.) Gaertn.	Ulezi, Uwimbi
	Tef	*Eragrostis tef* (Zuccagni) Trotter	Shairi
	Barley (2-row)	*Hordeum distichon* L.	
	Pearl millet	*Pennisetum glaucum* (L.) R.Br.	Uwele, Mtama
	Asian rice	*Oryza sativa* L.	Mpunga/Mchele/
	African rice	*Oryza glabberima* Steud.	Wali
	Sugar cane	*Saccharum officinarum* L.	
	Foxtail bristlegrass	*Setaria italica* L. Beauv.	Mua
	Sorghum	*Sorghum bicolor* (L.) Moench	Mtama
	Bread wheat	*Triticum aestivum* L.	Ngano
	Emmer wheat	*Triticum dicoccon* Schrank	
	Maize	*Zea mays* L.	Mhindi
Polygonaceae	Sorrel	*Rumex acetosa* L.	
Portulacaceae	Purselane	*Portulaca oleracea* L.	
Rhamnaceae	Jujube	*Ziziphus jujuba* (L.) Lam.	Mkunazi
Rhizophoraceae	Black mangrove	*Bruguiera gymnorrhiza* (L.) Lam.	Muia, Mshinzi,
	Yellow mangrove	*Ceriops tagal* C.B. Rob	Mchofi, Mkifu
	Red mangrove	*Rhizophora mucronata* Lam.	Mkandaa Mkoko, Mkaka
Rosaceae	Raspberry/blackberry	*Rubus* sp.	
Rubiaceae	Coffee	*Coffea* ssp.	Mkahawa
	Horseradish tree	*Moringa oleifera* Lam.	Mronge
Rutaceae	Lime	*Citrus aurantifolia*	Mdimu
	Lemon	*C. limon* (L.) Burm. f.	Mlimau
	Pomelo	*C. maxima* (Burm f.) Merr.	Mbalungi
	Sweet orange	*C. sinensis* (L.) Osbeck	Mchungwa
Solanaceae	Tomato	*Lycopersicon esculentum* Mill. (syn. *Solanum esculentum* L.)	Mtungule, Mnyanya
	Eggplant	*Solanum melongena* L.	Mbilingani
	Potato	*Solanum tuberosum* L.	Mbatata, Kiazi-mviringo
Sonneratiaceae		*Sonneratia alba* J.E. Smith	Mlilani, Mpia, mpira
Sterculiaceae		*Heritiera littoralis* Dryand.	Mkungu, Msikundazi,
Theaceae	Tea	*Camellia sinensis* (L.) O. Kuntze	Mchai
Verbenaceae	White mangrove	*Avicennia marina* (Forsk.) Veirh.	Mchu, Mtu

Appendix B
Plant Specimens Collected

No.	Family	Genus	Species	Infraspecies	Author(s)	Coll. Date	Det. by, date
1	Fabaceae	*Acacia*	*auriculiformis*		A. Cunn. ex Benth.	30-Jun-02	SCW, 2004
2	Fabaceae	*Senna*	*occidentalis*		(L.) Link	1-Jul-02	REG, 2004
3	Lamiaceae	*Ocimum*	*gratissimum*		L.	1-Jul-02	REG, 2004
4	Rubiaceae	*Agathisanthemum*	*bojeri*	var. bojeri		1-Jul-02	SCW, 2004
6	Dioscoreaceae	*Dioscorea*	*sansibarensis*		Pax	1-Jul-02	SCW, 2004
9	Cyperaceae	*Cyperus*			L.	1-Jul-02	SCW, REG, 2004
10	Euphorbiaceae	*Phyllanthus*	*nummulariifolius*		Poir.	1-Jul-02	SCW, 2004
12	Malvaceae	*Sida*	*cordifolia*		L.	10-Jul-02	SCW, 2004
13	Poaceae	*Sacciolepis*	*curvata*		(L.) Chase	10-Jul-02	REG, 2004
14	Asteraceae	*Ethulia*	*angustifolia*		Bojer ex DC.	10-Jul-02	REG, 2004
19	Vitaceae	*Ampelocissus*	*africana*	var. africana		18-Jun-04	SCW, 2004
20	Rubiaceae	*Keetia*	*zanzibarica*	subsp. zanzibarica		18-Jun-04	SCW, REG, 2004
21	Poaceae	*Digitaria*	*argyrotricha*		(Andersson ex Peters) Chiov.	18-Jun-04	REG, 2004
22	Cyperaceae	*Cyperus*	*amabilis*		Vahl	18-Jun-04	REG, 2004
23	Campanulaceae	*Lobelia*	*fervens*		Thunb.	7-Jul-04	SCW, REG, 2004
24	Poaceae	*Sacciolepis*	*curvata*		(L.) Chase	7-Jul-04	REG, 2004
25	Lamiaceae	*Ocimum*	*americanum*		L.	7-Jul-04	REG, 2004
26	Poaceae	*Eragrostis*	*setulifera*		Pilg.	7-Jul-04	REG, 2004
27	Asteraceae	*Vernonia*	*cinerea*	var. cinerea		7-Jul-04	REG, 2004
28	Cyperaceae	*Cyperus*	*compressus*		L.	7-Jul-04	REG, 2004
29	Cyperaceae	*Cyperus*	*maderaspatanus*		Willd.	7-Jul-04	REG, 2004
30	Asteraceae	*Launaea*	*cornuta*		(Hochst. ex Oliv. & Hiern) C. Jeffrey	7-Jul-04	REG, 2004
31	Tiliaceae	*Triumfetta*	*rhomboidea*		Jacq.	7-Jul-04	SCW, 2004
32	Malvaceae	*Abutilon*	*guineense*		(Schumach.) Baker f. & Exell	11-Jul-04	SCW, 2004
33	Sterculiaceae	*Heritiera*	*littoralis*		Dryand.	13-Jul-04	SCW, 2004
34	Combretaceae	*Lumnitzera*	*racemosa*		Willd.	13-Jul-04	SCW, 2004
35	Meliaceae	*Xylocarpus*	*granatum*		J. Koenig	13-Jul-04	SCW, 2004

SCW = Sarah Walshaw; REG = Roy Gereau, Missouri Botanical Garden (MO). Specimens 1-14 curated at MO and in author's collections only; remainder of specimens curated at MO, National Herbarium of Tanzania, Zanzibar Archives Herbarium, and in author's collections. All specimens entered into TROPICOS database.

Appendix C
Radiocarbon Data

Table C.1. Radiocarbon Dates from northern Pemba

Laboratory and Number	Method	Site	Context	Material	Radiocarbon Age (B.P.)	Calibrated Date (A.D.)
Beta-202827!	AMS	Tumbe	Unit 56, Layer 4, Level 1	Pearl millet grain	1370±40	665±15
Beta-191691*	Conventional	Tumbe	Unit 56, Layer 4, Level 1	Charred wood	1310±30	690±20
Beta-191689*	Conventional	Tumbe	Unit 16, Layer 4, Level 6	Charred wood	1280±40	730±50
Beta-202825!	AMS	Tumbe	Unit 14, Layer 3, Level 1	Coconut	1180±40	845±55
Pretoria-18070#	AMS	Chwaka	Unit 7, Layer 4	Charred wood	1000±30	1019±11
Pretoria-8241#	Conventional	Chwaka	Unit 7, Layer 2	Charred wood	750±40	1274±17
Pretoria-18069#	AMS	Chwaka	Unit 7, Layer 2	Charred wood	750±30	1274±13
Beta-202826!	AMS	Chwaka	Op-2C Layer 12, Level 1, Feature 192	Charred cotton seed	890±40	1070±20, 1170±30
Pretoria-8219#	Conventional	Kaliwa	Unit 1, Layer 3	Charred wood	540±50	1412±15
Beta-191612*	Conventional	Tumbe	Unit 49, Layer 3, Level 2	Charred wood	230±30	1660±10

Calibrated dates given in 1 sigma range. ! Indicates data from this study. *Indicates unpublished radiocarbon data.
# Indicates data reported in Fleisher 2003:A.1, using calibrated date for northern hemisphere.

Appendix D. Archaeobotanical Data

TUMBE Period 1 (n=23)

UNIT	BAG #	LAYER	LEVEL	FEATURE/CONTEXT	VOLUME (L)	LF sieve size (mm)	HF sieve size (mm)	INIT VOL. mL	IN.WEIGHT (G)	WOOD	WOOD WT (g)	STEM	STEM WT (g)	PARENCHYMA	PAR. WT (g)	COCONUT	COCONUT WT (g)	INDET NUT	INDET NUT WT (g)	ELAEIS?	UNID	UNID WT	ROOT/TUBER	ROOT/TUBER WT (g)
14	277	3	1		20	0.5	2.0	300	35.31	355	5.17	1				21	0.74	284	3.71		2	0.02		
14	296	3	2	7	3	0.5	1.0	60	23.62	45	0.40					4	0.06	15	0.15					
16	298	3	1		20			500	111.17	126	1.90					27	0.48	19	0.23		36	0.31		
16	300	4	1		18	0.5	1.0	400	30.93	124	2.14	4	0.03			11	0.21	23	0.22		16	0.2		
16	334	4	4		18	0.5	2.0	150	13.64	8	0.13						0	7	0.01		4	0.04		
47	603	4	1		10	0.5	1.0	100	11.08	pres	0.00					1	0.02							
50	389	3	4		10	0.5	2.0	100	64.89	45	0.68					2	0.03				2	0.01		
56	466	4	1		12	0.5	2.0	100	16.83	158	3.65						0				13	0.13	1	0.01
60	1004	3	1		10	0.5	2.0	200	56.64	62	0.50			2	0.02	10	0.14	16	0.14		2	0.02	2	0.01
60	1026	4	1		10	mesh	2.0		6.70	13	0.10					3	0.05	6	0.09					
60	1027	4	1	100	0.6	mesh	1.4		4.36	4	0.04						0	1	0.01		2	0.01		
60	1028	5	1		14	mesh	2.0		20.49	21	0.16					2	0.01							
63B	1183	3	1		12	mesh	2.0	300	76.55	221	7.63					5	0.13	18	0.18		26	0.22		
63B	1186	3	2	108	1	mesh	1.0		16.63	3	0.01					2	0.01				16	0.01		
63B	1275	4	1	149	10	mesh	1.4		48.98	168	5.79					21	0.84	27	0.58		16	0.18	1	0.01
63D	1468	3	1		6			1000	460.22		409.81					1	0.02	36	0.82					
64	1353	5	1		14	mesh	1.4	250	26.63	114	1.84			4	0.07	1	0.03	8	0.15		14	0.15	3	0.01
64	1354	5	1,2	152	14	0.5	2.0	150	22.38	53	0.56			3	0.01	9	0.13	11	0.14		6	0.04		
64	1376	6	1		11	mesh	1.4	50	14.77	18	0.16					2	0.02				2	0.03		
64	1599	5	1	154	15	0.5	1.4	200	11.97	125	1.75			12	0.11	3	0.02	10	0.12		3	0.03		
64	1631	5	1	162	18	0.5	2.0	150	11.91	45	0.70			1	0.01	2	0.02	6	0.12		2	0.01		
OP4E	1331	3	1		8	0.5	2.0	200	38.14	118	2.60			5	0.02	4	0.09	11	0.06		7	0.01		
OP4E	1429	4	1		8	0.5	2.0	150	14.10	55	0.95					4	0.11	8	0.09	1	1	0.01		
TOTALS					**262.6**						**446.7**	**5**	**0.03**	**27**	**0.2**	**135**	**3.2**	**506**	**6.8**	**1**	**170**	**1.4**	**7**	**0.0**

TUMBE Period 1 continued

UNIT	BAG #	PEARL MILLET	PEARL MILLET CHAFF	FINGER MILLET	SORGHUM	LIKE PEARL MILLET	TRITICOID	RICE	RICE CHAFF	INDET MILLET	UNID POACEAE	VIGNA SP.	FABACEAE	COTTON	RUMEX	POLYGONACEAE	EUPHORBIACEAE	PORTULACACEAE	ARGEMONE-TYPE	SESAME	INDET SEEDS	TOTAL SEEDS	UNID SEED FRAGS
14	277	19	1			11		2	4	4				1	6						2	46	56
14	296	125	9	1		85	1								2							218	55
16	298				1		2		8							1		1				5	39
16	300								6														47
16	334					1								1					1			2	17
47	603																						7
50	389	1						1					1							1		4	35
56	466	18	2		1	5								1							3	28	
60	1004							1														1	13
60	1026																						
60	1027																						
60	1028																				6	6	
63B	1183	1																			2	3	19
63B	1186																						1
63B	1275	3										4					1				10	18	110
63D	1468																				5	5	17
64	1353	2												4								6	23
64	1354	1						0			1			2	1						6	11	65
64	1376																						26
64	1599																		1		5	6	22
64	1631	2	1						1												1	3	19
OP4E	1331	2			1	1						1	2	1								8	16
OP4E	1429								1			1		1	1							5	14
TOTALS		**174**	**13**	**1**	**3**	**103**	**3**	**4**	**20**	**4**	**1**	**6**	**3**	**11**	**10**	**1**	**1**	**1**	**2**	**1**	**40**	**375**	**601**

TUMBE Period 1 continued

UNIT	BAG #	INSECT	FUNGUS	BONE - MAMMAL	BONE - MAMMAL WT (g)	BONE - FISH	BONE - FISH WT (g)	CORAL	CORAL WT (g)	SHELL	SHELL WT (g)	POTTERY	POTTERY WT (g)	DAUB	DAUB WT (g)	SEDIMENT	SEDIMENT WT (g)	SLAG	GLASS	<2mm RESIDUE
14	277			8	0.48					1	0.01	4	0.05			5	0.03			17.24
14	296															8	0.12			21.61
16	298			5	0.04	4	0.03					1	1.42				2.36			89.27
16	300																			11.04
16	334																			6.73
47	603																			5.01
50	389			5	0.03											2	0.04	4	1	62.36
56	466			1	0.01									7	0.31	2	0.06			7.78
60	1004	1										45	1.98				1.95		1	45.73
60	1026																			4.09
60	1027											7	0.83			11	0.20			2.96
60	1028											28	1.43	20	0.63		0.97			14.25
63B	1183	1		1	0.01							2	2.20	15	1.29		0.12			56.57
63B	1186											10	0.52				1.37		1	13.61
63B	1275	1		1	0.01							6	0.91	9	0.16					37.7
63D	1468	1												9	1.78					34.46
64	1353	1																		12.32
64	1354	1		19	0.59	1	0.01					2	0.06				0.03			15.29
64	1376			1	0.03															11.96
64	1599	1		2	0.01							1	0.01	2	0.01					3.26
64	1631	1		6	0.15			1	0.01			4	0.11							5.85
OP4E	1331		1									16	0.73			1	0.01			29.06
OP4E	1429	1						1	0.04			11	0.64							8.97
TOTALS		**9**	**1**	**49**	**1.36**	**5**	**0.04**	**2**	**0.05**	**1**	**0.01**	**137**	**10.9**	**62**	**4.2**	**29**	**7.3**	**4**	**3**	

MAZRUI TUMBE

UNIT	BAG #	LAYER	LEVEL	FEATURE/CONTEXT	VOLUME (L)	LF sieve size (mm)	HF sieve size (mm)	INIT VOL.	IN.WEIGHT (G)	WOOD CT	WOOD WT	STEM CT	STEM WT	PAR. CT	PAR. WT	COCO CT.	COCOWT.	Cocoeae	NUT CT.	NUT WT.	UNID CT	UNID WT	ROOT/TUBER
13	221	2	2		18	REI	1.00	300	67	354	4.25	6	0.04	15	0.07	8	0.38		100	1.33	53	1	
16	281	2	1		16	REI	1.00	400	62.9	238	2.93					77	0.83		3	0.05	19	0	3
44	562	2	1		10	0.50	1.00	175	11.2	14	0.15					6	0.08		23	0.29	14	0	1
53	418	2	1		12	REI	1.00	300	42.6	222	3.13	5	0.02	4	0.01	1	0.14	2	62	0.60	14	0	
60	1003	2	1		4.0	mesh	2.00	100	12.5	77	0.69	7	0.01			1	0.01				5	0	7
63C	1208	2	3	145	1.0	mesh	1.00	10mL	4.89	107	2.06			2	0.08				1	0.01	2	0	
64	1315	1	3&4	151	6.0	0.50	2.00	400	127		23.5	36	0.20			16	0.68		6	0.06	4	0	
64	1317	2	1		10.0	0.50	2.00	300	29.1	321	3.11	1	0.01	8	0.04	14	0.18		30	0.33	16	0	4
64 BCD	1417	2	1		13.0	0.50	2.00	500	40.3	950	10.7			10	0.06	15	0.29		45	0.42	38	0	
OP4E	1289	2	1		10.0	0.50	1.40	600	43.1	168	1.54	1	0.02	24	0.16	17	0.13		16	0.24	6	0	
Totals					**100**				**441**	**2451**	**52.1**	**56**	**0.3**	**63**	**0.42**	**155**	**2.72**	**2**	**286**	**3.33**	**171**	**2**	**15**

MAZRUI TUMBE

UNIT	BAG #	P MILL CT	GRASS/PMIL	RICE CT	RICE CHAFF	INDET MILLET	INDET POACEAE	FABACEAE	COTTON	RUMEX	POLYGONACEAE	COMMELINACEAE	MOLLUGO	Portulacaceae	PORTULACACEAE	INDET SEEDS	UNID SEED FRAG	UNCARB	DUNG	FUNGUS	BONE-MAMMAL	BONE-MAMMAL(g)	CORAL (g)	ct. POTTERY	POTTERY(g)	ct. DAUB	DAUB (g)	stone/sed	stone/sed (g)	<2mm RESIDUE
13	221	2													1	7	29	6.21						2	0.05					54.81
16	281	1		4	11				1		1			1		6	47	10.97			3	0								47.48
44	562								2								22	2.92			4	1.7								5.45
53	418				1		1			2		1				2	32	7.74		pres	1	0	0.05	2	0.01			4	0.1	30.37
60	1003		3	1	1											4	9	2.43	pres											8.99
63C	1208															1		0.21												2.55
64	1315															4	34	5.52	pres					12	0.72	26	0.4			84.12
64	1317	1			1		1		1				2		1	16	63	7.37	freq					2	0.02					16.44
64 BCD	1417	1	1		1					2			1?			0	23	12.83	pres											11.65
OP4E	1289		1	2	1			1	1						1	2	8	15.77	pres	pres	6	0.3	0.08	3	0.05			2	0.01	23.65
Totals		**5**	**5**	**7**	**16**	**1**	**1**	**1**	**5**	**4**	**1**	**1**	**3**	**1**	**3**	**42**	**267**	**71.97**			**14**	**2**	**0.13**	**21**	**0.85**	**26**	**0.4**	**6**	**0.11**	**285.51**

KIMIMBA Period 1 (n=17)

UNIT	BAG #	LAYER	LEVEL	VOLUME (L)	LF sieve size (mm)	HF sieve size (mm)	INIT VOL.	IN.WEIGHT (G)	WOOD CT	WOOD WT (g)	STEM CT	STEM WT (g)	PARENCHYMA CT	PARENCHYMA WT (g)	COCONUT CT	COCONUT WT (g)	COCOEAE CT	COCOEAE WT (g)	INDET NUT CT	INDET NUT WT (g)	ARECACEAE
3	102	2	1	20	0.50	2.00	200	18.01	291	3.97	2	0.01	16	0.12					12	0.11	
3	103	2	2	10	0.50	2.00	100	45.45		19.03	14	0.26	7	0.04					8	0.2	
3	104	2	3	21	0.50	2.00	200	76.79		25.9	5	0.02	107	0.91	4	0.02			7	0.07	
4	108	2	2	17	0.50	1.00	100	30.9	157	1.38	27	0.12	45	0.33	5	0.88			132	2.78	
4	107	3	1	10	0.50	1.00	100	10.15	43	0.3	120	0.78	7	0.03					83	0.56	
4	113	2	4	15	0.50	1.00	100	36.85	342	3.51	3	0.01	53	0.41	3	0.1			10	0.12	
7	133	3	1	17	0.50	2.00		22.37	44	0.55	1	0.01	2	0.01	1	0.01			1	0.01	
8	139	3	1	20	0.50	2.00	100	17.83	130	2.45			4	0.01					4	0.01	
8	141	4	1	22	0.50	2.00	150	87.37	73	0.98									1	0.01	
8b	161	2	2	12	0.50	2.00	75	30.69	78	0.92			1	0.01			1	0.01	4	0.02	
8c	175	2	2	11	REI	2.00		47.95	118	2.28	5	0.02	2	0.03	5	0.13			1	0.01	
9	146	2	2	6	REI	2.00	250	123.4		43.08	8	0.17							1	7.69	
10b	193	2	3	10	0.50	1.00	90	19.59	514	8.04											
11	182	2	2	15	0.50	2.00	200	18.44	144	2.03	1	0.01			2	0.01			30	0.52	
13	205	3	1	12			175	183.6	799	25.16	3	0.08	1	0.02					6	0.13	
14	272	3	1	3	0.50	1.00	75	31.68		12.01	2	0.09									
15	231	2	3	21	0.50	2.00	700	205.3		129.6	22	0.13	14	0.04	21	0.32			46	1.94	1
TOTALS				**242**						**281.2**	**213**	**1.71**	**259**	**1.96**	**41**	**1.47**	**1**	**0.01**	**346**	**14.2**	**1**

KIMIMBA Period 1 (n=17)

UNIT	BAG #	UNID CT	UNID WT (g)	ROOT/TUBER CT	ROOT/TUBER WT (g)	UNCARBONIZED	PEARL MILLET ? CT	INDET POACEAE	FABACEAE	MACROTYLOMA?	COTTON	FIG	RUMEX	RUBUS SP.	CUCURBITACEAE	AMARANTH	UNID SEED 'A'	UNID SEED 'B'	UNID SEED 'E'	UNID SEED 'D'	INDET SEEDS	TOTAL SEEDS	UNID SEED FRAGMENTS	UNID SEED FR WT (G)	UNCARBONIZED SEEDS
3	102	7	0.06	1	0	4.89															4	4	10		6
3	103	14	0.19			0.79			2					2							14	18	9		5
3	104	30	0.28			1.48			2							1		3			5	11	12		8
4	108	12	0.06	6	0.1	1.12			1		1		1		1						8	12	17		20
4	107			24	0.2	0.5								1							1	2	9		15
4	113	39	0.45	3	0.1	1.46			3	1											2	6	88	0.08	6
7	133		0.03	3	0	1.66															9	9	42	0.01	2
8	139	2	0.01			0.76																0	8		7
8	141	1	0.01			1.05					1											1	4		3
8b	161	2	0.01			1.36							1								2	3	4		
8c	175					0.98					2								1		1	4	17		9
9	146	14	0.19			0.08																0	5		3
10b	193	2	0.02			0.07	1					1									13	15	227	0.06	15
11	182	1	0.04			3.74		1												1	3	5	47	0.02	13
13	205	18	0.1			1.46								1				1			2	4	5		15
14	272	11	0.16			0.17																0			
15	231	51	1.67			16.01					1	1					3				15	20	54		11
TOTALS		**204**	**3.28**	**37**	**0.4**	**37.58**	**1**	**1**	**8**	**1**	**5**	**2**	**2**	**4**	**1**	**1**	**3**	**4**	**1**	**1**	**79**	**114**	**558**	**0.17**	**138**

KIMIMBA Period 1 (n=17)

UNIT	BAG #	INDET BUD	DUNG/INSECT	FUNGUS	CORAL CT	CORAL WT (g)	shell - ct	ctPOTTERY	wtPOTTERY	ct. DAUB	wt. DAUB	wtST/SED	ctST/SED	<2mm RESIDUE
3	102		pres											9.53
3	103		lots									0.02	5	24.61
3	104		lots											45.73
4	108		pres	pres			1	5	0.1			0.12		22.15
4	107		lots										1	7.67
4	113			pres										29.96
7	133			pres										19.85
8	139		pres	pres										14.55
8	141		pres	pres										84.9
8b	161		lots	pres										28.14
8c	175		pres	pres										44.32
9	146													71.69
10b	193													10.44
11	182	1								1	0.04	0.05	4	11.76
13	205		lots									0.15	15	155.2
14	272				9	0.08								19.11
15	231		lots	lots										55.85
TOTALS		**1**			**9**	**0.08**	**1**	**5**	**0.1**	**1**	**0.04**	**0.34**	**25**	

KALIWA Period 3 (n=11)

UNIT	BAG#	LAYER	LEVEL	VOLUME (L)	LF sieve size (mm)	HF sieve size (mm)	INITIAL VOL. (mL)	INITIAL WEIGHT (G)	WOOD CT	WOOD WT (g)	STEM CT	STEM WT (g)	PARENCHYMA CT	PARENCHYMA WT (g)	COCONUT CT	COCONUT WT (g)	ELAEIS?	INDET NUT CT	INDET NUT WT (g)	UNID NUT"C" CT	UNID NUT"C" WT (g)	UNID NUT "D" CT	UNID NUT "D" WT (g)
2-B	184	3	2	10	0.50	2.00	75	20.87	91	2.54	4	0.03	3	0.08	1	0.01	1	188	3.27				
2-F	228	1	3	10	0.50	2.00	300	21.74	131	1.44	1	0.01	6	0.03	3	0.1		38	0.5				
2-F	229	2	1	10	0.50	2.00	150	22.69	98	1.18	4	0.03	30	0.25	11	0.27		81	2.36				
3	169	2	1	12	0.50	1.40	300	51.8	98	1.41			42	0.26	1	0.04		65	0.91	2	0.60	2	0.26
3	171	3	1	10	0.50	2.00	200	19.39	149	2.31	4	0.01	1	0.01	6	0.12		94	1.07				
4	201	2	2	12	0.50	1.40	100	21.87	133	0.88	2	0.02	7	0.06	2	0.04	5	231	1.82				
4	203	3	1	8	mesh	1.40	75	34.98	65	0.6	1	0.01			1	0.06		51	0.7				
5	212	1	1	13	0.50	2.00	250	86.96	40	0.33	2	0.01	13	0.07	5	0.03		48	0.28				
5	215	1	4	8	0.50	2.00	150	46.83	194	2.44	8	0.04	6	0.06	4	0.08		565	8.07				
5	222	2	1	8	0.50	2.00	100	59.2	97	1.3	6	0.01			10	0.36	18	411	6.39				
5	223	2	2	10	0.50	2.00	150	43.29	128	2.83	12	0.04			14	0.72	5	1	7.91				
TOTALS				**111**				**429.62**	**1224**	**17.26**	**44**	**0.21**	**108**	**0.82**	**58**	**1.83**	**29**	**1773**	**33.28**	**2**	**0.60**	**2**	**0.26**

KALIWA Period 3 (n=11) continued

UNIT	BAG #	UNID CT	UNID WT (g)	UNCARBONIZED (g)	PEARL MILLET	FINGER MILLET	RICE	RICE CHAFF	RICE EMBRYO	UNID MILLET	UNID MILLET CHAFF	UNID POACEAE	FABACEAE	COTTON	SIDA?	POLYGONACEAE	CARYOPHYLLACEAE	MOLLUGO	INDET SEEDS	TOTAL SEEDS	UNID SEED FRAGS	UNCARBONIZED SEEDS
2-B	184			1.1			1	1						3					1	5	7	11
2-F	228	6	0.12	8.5		1		3						1	1				2	5	4	1
2-F	229	82	0.88	6.3			2	5						1					0	3	32	87
3	169	30	0.56	11			1	1						2		1		1	0	5	29	71
3	171	11	0.08	7.8				6	1		1			2					2	4	14	19
4	201	12	0.15	1.4			7	27	1					3					4	14	25	36
4	203			1.7			2	19						1						3	7	7
5	212	2	0.01	6	3		1		2					1						5	7	1
5	215	4	0.02	3.3			2	4		1		1		5					0	9	18	2
5	222	11	0.15	0.7			2	3						11			2		3	18	30	5
5	223	21	0.13	2.2	1		3	3	3				1	1		1	1		2	10	21	25
TOTALS	**179**	**2.10**	**49.8**	**4**	**1**	**21**	**72**	**7**	**1**	**1**	**1**	**1**	**31**	**1**	**2**	**3**	**1**	**14**	**81**	**194**	**265**	

KALIWA Period 3 (n=11) continued

UNIT	BAG #	DUNG/INSECT	FUNGUS	BONE-MAMMAL CT	BONE- MAMMAL WT (g)	BONE-FISH CT	BONE- FISH WT (g)	CORAL CT	CORAL WT (g)	SHELL CT	SHELL WT (g)	POTTERY CT	POTTERY WT (G)	DAUB CT	DAUB WT (g)	STONE/SED WT (g)	SLAG CT	<2mm RESIDUE
2-B	184			2	0.01	4	0.03	1	0.02			3	0.4			1.16		11.09
2-F	228			1	0.02	1	0.01	4	0.06			4	0.1			0.01		11.85
2-F	229					1	0.01					1	0	11	0.11	0.03	7	9.34
3	169					1	0.01					9	0.6					32.3
3	171	pres										6	0.2			0.01		7.4
4	201	pres						7	0.14			3	0.1			0.03		15.83
4	203			9	0.09			5	0.04			3	0.2					31.34
5	212	pres		4	0.05	2	0.03			1	0.03	34	0.9	17	0.38	0.17		77.43
5	215	pres		10	0.1	10	0.24									4.85		11.23
5	222	pres		12	0.14	139	2.37					3	0	1	0.05	0.53		46.59
5	223	pres	pres	5	0.44	204	2.97					1	0			0.01		21.62
TOTALS				**43.0**	**0.85**	**362**	**5.67**	**17.0**	**0.26**	**1.0**	**0.03**	**67**	**2.51**	**29**	**0.54**	**6.80**	**7**	

Appendix E
Archaeobotanical data from Chwaka

CHWAKA Periods 2 & 3 (n=34)

UNIT	BAG#	LAYER	LEVEL	FEATURE/CONTEXT	VOLUME (L)	LF sieve size (mm)	HF sieve size (mm)	INITIAL VOL.(mL)	INITIAL WEIGHT (g)	WOOD CT	WOOD WT (g)	STEM CT	STEM WT (g)	PARENCHYMA CT	PARENCHYMA WT (g)
11	226	2	5		20	0.5	2.0	100	87.07	468	7.36	25	0.18	6	0.04
11	241	3	2		20	0.5	2.0	300	175.10	1073	21.32	37	0.27	5	0.09
11	265	4	4		16	0.5	1.0	60	60.70	307	4.95	14	0.12	5	0.02
12	230	3	1		20	0.5	2.0	125	62.95	405	6.99	13	0.07	19	0.21
19	446	3	1		10	0.5	1.0	100	73.01	241	4.50	27	0.21	10	0.17
22	442	3	1		8	0.5	1.0	200	96.22		7.32	70	0.55		
26	528	3	1		12	0.5	1.0	125	12.47	50	0.38	2	0.01	5	0.05
27	547	2	3		12			100	103.00	300	6.94	47	0.38	45	0.4
27	558	3	4		13	0.5	1.0		69.96	720	12.49	37	0.25		
38	650	2	6	20	2	0.5	1.0	20	2.83	20	0.63				
39	625	2	1		12	0.5	1.0	140	16.60	81	0.78			3	0.04
Op-1A	1453	12	1		12	0.5	1.4	50	19.59	139	1.55	22	0.01		0
Op-1B	1067	1	4		10	0.5	1.4	150	54.70	799	8.46			37	0.24
Op-1B	1069	3	1		9	0.5	1.0	150	39.92		7.69	9	0.04	32	0.35
Op-1C	1174	2	1		9	0.5	1.4	100	52.41	653	6.59	1	0.01	4	0.05
Op-1C	1482	8	1		8	0.5	2.0	30	5.26	91	0.58	1	0.01		
Op-1F	1602	5	1		8	0.5	1.4	50	13.29	89	1.02	28	0.13	19	0.08
Op-2A	1322	4	1		9	0.5	2.0	100	28.41	76	3.25	2	0.01		
Op-2B	1096	2	1		8	mesh	1.4	100	38.33	28	2.00				
Op-2B	1144	3	1		8	mesh	1.4	150	43.53	51	1.15				
Op-2B	1166	4	1		10	mesh	1.4	150	34.27	120	3.93	14	0.12		
Op-2B	1219	5	1		9	0.5	2.0	50	20.57	67	1.64	4	0.01	2	0.01
Op-2B	1225	6	1		10	0.5	1.4	200	93.71	655	25.45	67	0.16	9	0.04
Op-2B	1266	7	1		9	mesh	1.4	200	102.62	356	9.19	47	0.39	12	0.14
Op-2B	1281	7	2		10	0.5	1.4	350	155.58		21.46			4	0.06
Op-2B	1283	8	1	143	1	0.5	1.0	40	19.36	557	6.66			3	0.03
Op-2B	1284	8	1		8	0.5	2.0	75	33.67	154	2.08	3	0.01		
Op-2C	1285	4	1		8	0.5	2.0	25	5.74	42	1.40	3	0.01	2	0.01
Op-2C	1557	10	3	188	0.5	0.5	1.0	75	28.04	508	17.19			19	0.15
Op-2C	1611	12	1	192	6	0.5	1.4		64.40	832	10.48	1	0.03		
Op-2D	1165	4	1		12	0.5	2.0	250	40.33	123	2.21	1	0.01		
Op-2D	1573	14	1	171	10	0.5	1.4	150	46.54	317	2.98	11	0.06	61	0.33
Op-2E	1218	4	1	141	8	0.5	1.4	350	25.35	33	0.49	1	0.01	9	0.06
Op-5	1385	28			14	0.5	2.0	250	127.98	976	13.34	5	0.06	12	0.13
TOTALS					**342**						**224.45**	**492**	**3.121**	**323**	**2.7**

UNIT	BAG#	COCONUT CT	COCONUT WT (g)	ELAEIS? CT	ELAEIS? WT (g)	INDET NUT CT	INDET NUT WT	UNID BOT CT	UNID BOT WT (g)	ROOT/TUBER CT	ROOT/TUBER WT (g)	UNCARBONIZED (g)	PEARL MILLET	FINGER MILLET	SORGHUM	SORGHUM CHAFF	LIKE PEARL MILLET
11	226	109	2.43	7	0.08	328	5.63		4.64	1	0.01	0.79					1
11	241	81	3.29	7	0.14	1226	16.92	11	0.06			0.35					
11	265	109	1.97			242	2.77	16	0.12			0.08					
12	230	109	2.07	3	0.03	747	9.43	24	0.13			0.89					
19	446	54	1.76	9	0.16	458	5.59		0			0.07					
22	442	50	2.27	2			15.92	1	0.01	1	0.14	0.35				1	
26	528	5	0.22			21	0.25	17	0.14	2	0.05	1.94	1				
27	547	72	1.73			161	2.79		1.14			1.12					
27	558	85	3.08	3	0.08	484	5.84		0.05			0.98					
38	650					4	0.02	19	0.07			0.26					
39	625	34	1.13			46	0.61	11	0.08			3.5					
Op-1A	1453	12	0.33	4		337	3.54	6	0.01			0.07					
Op-1B	1067	39	1.26			148	3.32	10	0.08			1.43	2				
Op-1B	1069	58	1.37			138	1.46	22	0.16			0.34	2				6
Op-1C	1174	33	1.27			72	1.08	36	0.5			0.27	4				
Op-1C	1482	4	0.04	3	0.03	105	0.74			1	0.01	0.12					
Op-1F	1602	20	0.53	2		192	1.68	5	0.07			0.06					
Op-2A	1322					2	0.03	3	0.04			4.15					
Op-2B	1096	pres				4	0.06	1	0.01			3.19					
Op-2B	1144	3	0.11			12	0.12	5	0.1			3.68					
Op-2B	1166	2	0.08			3	0.01					2.28					
Op-2B	1219	2	0.02			13	0.14					1.24					
Op-2B	1225	4	0.19			6	0.06	30	0.19			0.53	1	1	1		
Op-2B	1266	6	0.13	8	0.11	50	0.76	30	0.13			0.87			2		
Op-2B	1281	70	2.20			49	0.45	98	0.41			0.7					3
Op-2B	1283					2	0.03	4	0.01			0.15					
Op-2B	1284	3	0.12			1	0.01	2	0.02			0.68					
Op-2C	1285	1	0.01			12	0.11					1.31					
Op-2C	1557	6	0.29			41	0.31	4	0.04			0.02					
Op-2C	1611	75	1.02			606	8.58	1	0.42	1	0.04	0.38					
Op-2D	1165	12	0.44			117	1.63	4	0.03			5.02					
Op-2D	1573	34	0.52			284	4.3	1	0.3			2.21					
Op-2E	1218	2	0.01			8	0.05	4	0.02			5.91					
Op-5	1385	75	3.00	20	0.37	729	7.18	20	0.12			3.56			2		
TOTALS		**1169**	**32.89**	**68**			**101.42**		**9.1**	**6**		**48.5**	**10**	**1**	**5**	**1**	**10**

UNIT	BAG #	TRITICEAE	RICE	RICE WT	RICE EMBRYO	RICE CHAFF	RICE CHAFF WT	DIGITARIA	INDET MILLET	INDET MILLET CHAFF	INDET POACEAE	IND GR CH	VIGNA SP.	V. RADIATA	PISUM SP.	LIKE PISUM	FABACEAE	COTTON	COTTON WT	POPPY	FIG
11	226		2	0.01	2	45					1			1	1			12	0.05		
11	241		6	0.03	1	11												6	0.08		2
11	265					4											1	14	0.07		
12	230		5		9	75				2								10	0.02		
19	446					6												6			
22	442		3		3	15											1	20			
26	528	1	5	0.02		38		1				6						2	0		
27	547		2			10											2	5	0.03		
27	558		1			3									1			27	0.16		
38	650		1			1												1			
39	625		1			23								1							
Op-1A	1453		1			47			1									9	0.03		
Op-1B	1067		1		2	5					2						1				
Op-1B	1069		6		1	111			1	1			1					1			
Op-1C	1174		1		2	27												1			
Op-1C	1482		0		1	5												4			
Op-1F	1602		0		3	12												5			
Op-2A	1322		24			7															
Op-2B	1096		31			10												1			
Op-2B	1144		381	3.18	2	78												1			
Op-2B	1166		9			3															
Op-2B	1219		191	1.64		71												1			
Op-2B	1225		3929	35.36	1	979					1			1	16	23		2			
Op-2B	1266		4009	35.26	1	1174	0.21				1				17			4			
Op-2B	1281		6000	60.65		1750	0.35								7	8	1	5			
Op-2B	1283		102	0.92		17												1			
Op-2B	1284		388	3.34		113												1			
Op-2C	1285		11			4							1					1			
Op-2C	1557				1	2											1	1			
Op-2C	1611		16		79	523	0.07											404	6.43		
Op-2D	1165		4		1	1												2			
Op-2D	1573					50								1				7	0.04		
Op-2E	1218		13	0.01		2												1			
Op-5	1385		3	0.01	3	121									1			36		1	
TOTALS		**1**	**15146**		**112**	**5343**		**1**	**2**	**3**	**5**	**6**	**2**	**4**	**43**	**31**	**7**	**591**		**1**	**2**

UNIT	BAG #	RUMEX	POLYGONACEAE	RANUNCULACEAE	CARYOPHYLLACEAE	PORTULACACEAE	CHENOPODIACEAE	INDET SEEDS	TOTAL SEEDS	UNID SEED FRAGMENTS	UNID SEED FR WT (g)	UNID NONBOT	UNCARBONIZED SEEDS	DUNG/INSECT	FUNGUS	BONE-MAMMAL CT	BONE-MAMMAL WT (g)	BONE-FISH CT	BONE-FISH WT (g)
11	226							1	19		0.34		28						
11	241								24	10			6			67	2.47	305	5.09
11	265	5			1				21	20	0.03		14						4.42
12	230							3	18		0.07		6					264	3.57
19	446	1			3				10	8			13	pres		24	1.32	248	3.04
22	442								24	14			4	pres			2.24		11.41
26	528		1	1	1			2	14	208	0.14		43	pres		4	0.09	2	0.03
27	547							4	13		0.22		100					37	0.38
27	558	1	1					1	32		0.21		15			3	0.08		2.16
38	650								2	26	0.01		3					1	0.01
39	625					1		1	4	62	0.04		101			4	0.08		
Op-1A	1453		1					3	15	33	0.01		5						0.48
Op-1B	1067							2	8	28			11	pres	pres	14	0.44		3.85
Op-1B	1069							1	23	124	0.15		22	pres	pres	6	0.1		2.64
Op-1C	1174								6	3		6	11	pres		57	1.26		8.66
Op-1C	1482								4	3			5					6	0.05
Op-1F	1602							1	6	42			12	pres				15	0.26
Op-2A	1322								24	4			11						
Op-2B	1096								9	14			17			1	0.05	1	0.01
Op-2B	1144							1	383	4			26					5	0.04
Op-2B	1166							1	10				2						
Op-2B	1219								192	4			10	pres					
Op-2B	1225						1	3	3956	21			7	pres					
Op-2B	1266							4	4037	7			11			6	0.08	3	0.03
Op-2B	1281							8	6032		0.03		8			8	0.23		
Op-2B	1283							4	106	33									
Op-2B	1284							3	392				16						
Op-2C	1285								13	1			4	pres	pres			1	0.01
Op-2C	1557								2	5	0.01					1	0.01	1	0.01
Op-2C	1611							1	421	89	0.05								
Op-2D	1165								6	6			30	pres		21	0.35	9	0.09
Op-2D	1573							4	11	45			16	pres					4.8
Op-2E	1218								14	14		4	100						
Op-5	1385							2	44	17		2	15	pres		96	3.28		4.84
TOTALS		**7**	**3**	**1**	**5**	**1**	**1**	**50**	**15895**	**845**		**12**	**672**			**312**	**12.08**	**898**	**55.88**

CHWAKA Periods 2 & 3 (n=34)

UNIT	BAG #	SHELL CT	SHELL WT (g)	CORAL CT	CORAL WT (g)	POTTERY CT	POTTERY WT (g)	DAUB CT	DAUB WT (g)	CORAL RAG CT	CORAL RAG WT (g)	STONE/SED CT	STONE/SED WT (g)	SLAG	GLASS	BEADS	<2mm RESIDUE
11	226												2.51				56.79
11	241					6	0.07					86	2.05				120.73
11	265			18	0.12	1	0.02					15	0.80	8	1		45.03
12	230											3	0.04				38.69
19	446			1		2	0.05										56.14
22	442			40	0.80	8	0.14					5	0.10				55.01
26	528					3	0.04					8	0.08				8.79
27	547												3.12				84.54
27	558			4	0.13	2	0.02						0.43				43.82
38	650							1	0.01			8	0.09				1.70
39	625											2	0.09				11.43
Op-1A	1453															1	11.09
Op-1B	1067			25	0.69	16	0.99						0.96				29.64
Op-1B	1069			18	0.35	2	0.02	26	0.40								21.69
Op-1C	1174	1	0.05	25	0.56	18	1.08					27	0.44				28.94
Op-1C	1482			9	0.10	6	0.12					7	0.06				2.91
Op-1F	1602				0.61	2	0.05						0.26				7.66
Op-2A	1322					4	0.67						3.02				16.75
Op-2B	1096					1	0.01						10.10				22.54
Op-2B	1144					5	0.39						3.30				31.21
Op-2B	1166												5.17				21.68
Op-2B	1219					10	0.27						0.93				14.21
Op-2B	1225					4	0.2										27.38
Op-2B	1266			5	0.05	53	1.13						0.36				44.38
Op-2B	1281					5	1.1		5.25								41.94
Op-2B	1283																9.77
Op-2B	1284					4	0.16						1.50				25.00
Op-2C	1285			1	0.29							3	0.30				1.97
Op-2C	1557					32	1.08					3	0.07				6.95
Op-2C	1611			1	0.01								0.60				26.95
Op-2D	1165			1	9.46	7	0.2						1.18				17.35
Op-2D	1573	1	0.01					1	0.05			3	0.08				25.64
Op-2E	1218			1	5.75												12.17
Op-5	1385	29	0.48	87	3.45	95	3.02			50	0.84	35	0.74				75.27
TOTALS		**31**	**0.54**	**236**	**22.37**	**286**	**10.83**		**5.71**	**50**	**0.84**	**205**	**38.38**	**8**	**1**	**1**	

References Cited

[APG], Angiosperm Phylogeny Group
1998 An ordinal classification for the families of flowering plants. *Annals of the Missouri Botanical Garden* 85:531-553.
2003 An update of the Angiosperm Phylogeny Group classification for the orders and families of flowering plants: APG II. *Botanical Journal of the Linnaean Society* 141:399-436.

Abdel-Magid, A.
1989 *Plant domestication in the middle Nile basin: an archaeobotanical case study*. British Archaeological Reports, Oxford.

Abungu, G. H. O.
1994-95 Agriculture and settlement formation along the East African Coast. In *The Growth of Farming Communities in Africa South of the Equator*, edited by J. E. G. Sutton, pp. 248-256, Azania Special Volume XXIX-XXX.
1998 City states of the East African coast and their maritime contacts. In *Transformations in Africa: essays on Africa's later past*, edited by G. Connah, pp. 204-218. Leicester University Press, London & Washington.

Abungu, G. H. O. and H. Mutoro
1993 Coast interior settlements and social relations in the Kenya coastal hinterland. In *The archaeology of Africa: Foods, metals and towns*, edited by T. Shaw, P. J. J. Sinclair, B. Andah and A. Okpoko, pp. 694-704. Routledge, London.

Acland, J. D.
1971 *East African Crops*. FAO, Longman, Rome.

Adams, R. M.
1966 *The Evolution of Urban Society: Early Mesopotamia and Prehispanic Mexico*. Aldine, Chicago.
1981 *Heartland of Cities*. University of Chicago Press, Chicago.

Allen, J. de V.
1993 *Swahili Origins*. Ohio University Press, Athens.

Amblard, S. and J. Pernès
1989 The identification of cultivated pearl millet (*Pennisetum*) amongst plant impressions on pottery from Oued Chebbi (Dar Oualata, Mauritania). *African Archaeological Review* 7:117-126.

Ambrose, S. H.
1984 Excavations at Deloraine, Rongai, 1978. *Azania* XIX:79-104.

Anderberg, A.-L.
1994 *Atlas of seeds and small fruits of Northwestern-European plant species with morphological descriptions: Part 4 Resedaceae - Umbelliferae*. Swedish Museum of Natural History, Stockholm.

Anderson, E.
1997[1952] *Plants, man and life*. Missouri Botanical Garden, St Louis, MO.

Bedigian, D.
2003 Sesame in Africa: origins and dispersals. In *Food, fuel and fields: progress in African Archaeobotany*, edited by K. Neumann, A. Butler and S. Kahlheber, pp. 17-36. Heinrich Barth Institut, Koln.
2004 History and lore of sesame in Southwest Asia. *Economic Botany* 58(3):329-353.

Beentje, H. J.
1990 *Botanical assessment of Ngezi forest, Pemba*. Zanzibar Forestry Development Project of FINNIDA and the Finnish National Board of Forestry.

Berggren, G.
1969 *Atlas of seeds and small fruits of Northwest-European plant species with morphological descriptions: Part 2 Cyperaceae*. Swedish Natural Science Research Council, Stockholm.
1981 *Atlas of seeds and small fruits of Northwest-European plant species with morphological descriptions: Part 3 Salicaceae-Cruciferae*. Swedish Museum of Natural History, Stockholm.

Blench, R.
1994 The ethnographic evidence for long-distance contacts between Oceania and East Africa. In *The Indian Ocean in antiquity*, edited by J. Reade, pp. 417-438. Kegan Paul International, London.

Boardman, S.
1999 The agricultural foundation of the Aksumite Empire: an interim report. In *The Exploitation of Ancient Plant Resources in Ancient Africa*, edited by M. van der Veen, pp. 137-147. Klewer Academic/Plenum, New York.
2000 Archaeobotany. In *Archaeology at Aksum, Ethiopia, 1993-97*, edited by D. W. Phillipson, pp. 127, 156, 217, 363-368, 412-414. 2 vols. British Institute in East Africa and the Society of Antiquaries of London, London.

Boserup, E.
1965 *The conditions of agricultural growth; the economics of agrarian change under population pressure*. Aldine Pub. Co., Chicago.

Brain, C. K.
1974 Human food remains from the Iron Age at Zimbabwe. *South African Journal of Science* 70:303-309.

Brumfiel, E. and T. K. Earle
1987 Specialization, exchange, and complex societies: An introduction. In *Specialization, exchange, and complex societies*, edited by E. Brumfiel and T. K. Earle, pp. 1-9. Cambridge University Press, Cambridge.

Brunken, J. N.
1977 A Systematic Study of Pennisetum sect. Pennisetum (Gramineae). *American Journal of Botany* 64(2):161-176.

Buchanan, L. A. C.
1931 *The Ancient monuments of Pemba*. The Zanzibar Museum, Zanzibar.

Butler, A. and A. C. D'Andrea
2000 Farming and famine: subsistence strategies in highland Ethiopia. In *The archaeology of drylands: living at the margin*, edited by G. Barker and D. Gilbertson, pp. 180-197. Routledge, London.

Cain, C. R.
2000 *Animals at Axum: initial zooarchaeological research into the later prehistory of the northern Ethiopian highlands.* Unpublished PhD dissertation,Washington University in St Louis.
2005 Towns, politics, and meat in ancient north-eastern Pemba Island, Tanzania. In *Pan-African archaeological congress for prehistory and related studies*, Gaborone, Botswana.

Cappers, R. T. J.
2003 Exotic imports of the Roman Empire: an exploratory study of potential vegetal products from Asia. In *Food, fuel and fields: progress in african archaeobotany*, edited by K. Neumann, A. Butler and S. Kahlheber, pp. 197-206. Heinrich Barth Institut, Koln.

Chami, F.
1994 *The Tanzanian Coast in the First Millennium AD: An archaeology of the iron working, farming communities* Studies in African Archaeology 7. Societas Archaeologica Uppsaliensis, Uppsala.
1998 A review of Swahili Archaeology. *African Archaeological Review* 15(3):199-218.

Champion, L. and D. Fuller
2015 Cereal farming and early urbanim in northern Benin: archaeobotanical results from twelve sites. In *Società dei Naturalisti e Matematici di Modena*, Suppl. Vol CXLVI Proceedings of the IWAA8 pp. 171-173.

Chang, T.-T.
1976 Rice. In *Evolution of Crop Plants*, edited by N. W. Simmonds, pp. 98-104. Longman, London.

Childe, V. G.
1950 The urban revolution. *Town Planning Review* 21:2-17.

Chittick, N.
1971 The coast of East Africa. In *The African Iron Age*, edited by P. L. Shinnie, pp. 108-141. Clarendon Press, Oxford.
1974 *Kilwa: an Islamic trading city on the East African Coast.* Memoir No. 5 1: History and Archaeology. 2 vols. British Institute in East Africa, Nairobi.
1984 *Manda: Excavations at an Island Port on the Kenya Coast.* Memoir 9, British Institute in Eastern Africa, Nairobi.

Colson, E.
1979 In good years and in bad: food strategies of self-reliant societies. *Journal of Anthropological Research* 35:18-20.

Connah, G.
1975 *The Archaeology of Benin.* Clarendon Press, Oxford.
1985 Agricultural intensification and sedentism in the firki of N.E. Nigeria. In *Prehistoric intensive agriculture in the tropics*, edited by I. S. Farrington, pp. 765-785. BAR International Series 232, Oxford.
2001 *African Civilizations: an archaeological perspective.* 2nd ed. Cambridge University Press, Cambridge.

Coquery-Vidrovitch, C.
1971 *Research into an African Mode of Production.* United Nations African Institute for Economic Development and Planning, Dakar.
1997 Research on an African Mode of Production. In *Perspectives on Africa: a reader in culture, history, and representation*, edited by R. R. Grinker and C. B. Steiner, pp. 129-141. Blackwell, Cambridge, MA.

Costin, C. L.
1991 Craft specialization: Issues in defining, documenting, and explaining the organization of production. *Archaeological Method and Theory* 3:1-56.

Cotton, C. M.
1996 *Ethnobotany: principles and applications.* John Wiley & Sons, Chichester; New York.

Craster, J. E. E.
1913 *Pemba, the spice island of Zanzibar*. Unwin.

Crawford, G. W. and C. Shen
1998 The origins of rice agriculture: recent progress in East Asia. *Antiquity* 72(278):858-866.

D'Andrea, A. C., M. Klee and J. Casey
2001 Archaeological evidence for pearl millet (*Pennisetum glaucum*) in sub-Saharan West Africa. *Antiquity* 75:341-348.

D'Andrea, A. C., A. Logan and D. Watson
2006 Oil palm and prehistoric subsistence in West Africa. *Journal of African Archaeology* 4 (2):195-222

D'Andrea, A. C., D. E. Lyons, Mitiku Haile and E. A. Butler
1999 Ethnoarchaeological approaches to the study of prehistoric agriculture in the highlands of Ethiopia. In *The Exploitation of Ancient Plant Resources in Ancient Africa*, edited by M. van der Veen, pp. 101-122. Klewer Academic/Plenum, New York.

D'Andrea, A. C. and Mitiku Haile
2002 Traditional emmer processing in highland Ethiopia. *Journal of Ethnobiology* 22(2):179-217.

D'Andrea, A. C. and J. Casey
2002 Pearl millet and Kintampo subsistence. *African Archaeological Review* 19:147-173.

Damerow, P.
1996 Production and social status as documented in proto-cuneiform texts. In *Food and the Status Quest: an Interdisciplinary Perspective*, edited by Polly Weissner and Wulf Schieffenhövel, pp. 149-169. Berghahn Books, Oxford.

Datoo, B. A.
1970 Rhapta: the location and importance of East Africa's port. *Azania* 5:66-75.

Dennell, R. W.
1974 Prehistoric crop processing activities. *Journal of Archaeological Science* 1:275-284.
1976 The economic importance of plant resources represented on archaeological sites. *Journal of Archaeological Science* 3:229-247.

Dennis, J. V. and C. R. Gunn
1971 Case against Trans-Pacific dispersal of the coconut by ocean currents. *Economic Botany* 25:407-413.

Dharani, N.
2002 *Field Guide to Common Trees and Shrubs of East Africa.* Struik, Cape Town.

Dietler, M.
1996 Feasts and commensal politics in the political economy: tood, power, and status in prehistoric Europe. In *Food and the Status Quest: an interdisciplinary perspective*, edited by P. Wiessner and W. Schiefenhövel, pp. 87-125. Berghahn Books, Oxford.
2001 Theorizing the Feast: rituals of consumption, commensal politics, and power in African contexts. In *Feasts: Archaeological and ethnographic perspectives on food, politics, and power*, edited by M. Dietler and B. Hayden, pp. 65-114. Smithsonian Institution Press, Washington.

Dietler, M. and B. Hayden (editors)
2001 *Feasts: archaeological and ethnographic perspectives on food, politics, and power.* Smithsonian Institution Press, Washington.

Dietler, M. and I. Herbich
2001 Feasts and Labor Mobilization: dissecting a fundamental economic practice. In *Feasts: Archaeological and ethnographic perspectives on food, politics and power*, edited by M. Dietler and B. Hayden, pp. 240-263. Smithosonian Institution Press, Washington.

Donley-Reid, L.
1990 A Structuring Structure: The Swahili House. In *Domestic Architecture and the Use of Space*, edited by S. Kent, pp. 114-126. Cambridge University Press, Cambridge.

Filippi, M. L. and M. R. Talbot
2005 The palaeolimnology of northern Lake Malawi over the last 25 ka based upon the elemental and stable isotopic composition of sedimentary organic matter. *Quaternary Science Reviews* 24(10-11):1303-1328.

Fitzgerald, W.
1945 *Africa: a social, economic and political geography of its major regions.* 5th ed. Metheun & Co. Ltd., London.

Fitzgerald, W., W.A.
1970[1898] *Travels in the coastlands of British East Africa and the islands of Zanzibar and Pemba: their agricultural resources and general characteristics.* Dawsons of Pall Mall, Folkestone & London.

Fleisher, J.
2001 Archaeological survey and excavations in Northern Pemba Island, Tanzania, 1999-2000. *Nyame Akuma* 56:36-43.
2003 *Viewing stonetowns from the countryside: an archaeological approach to Swahili regional systems AD 800-1500.* Unpublished PhD Dissertation, University of Virginia.2010 Rituals of consumption and the politics of feasting on the Eastern African coast, AD 700-1500. *Journal of World Prehistory* 23(4):195-217.

Fleisher, J., P. Lane, A. LaViolette, M. Horton, E. Pollard, E. Quintana Morales, T. Vernet, A. Christie, and S. Wynne-Jones
2015 When Did the Swahili Become Maritime? *American Anthropologist* 117(1):110-115

Fleisher, J. and A. LaViolette
1999a Elusive Wattle-and-Daub: Finding the Hidden Majority in the Archaeology of the Swahili. *Azania* 34:87-108.
1999b The Recovery of Swahili Settlements in the Absence of Stone Architecture: Two Preliminary Surveys from Pemba Island, Tanzania. *Nyame Akuma* 52:64-78.
2013 The early Swahili trade village of Tumbe, Pemba Island, Tanzania, AD 600-950. *Antiquity* 87 (2013): 1151-1168

Fletcher, R.
1998 African urbanism: scale, mobility and transformations. In *Transformations in Africa: Essays on Africa's later past*, edited by G. Connah, pp. 104-138. Leicester University Press, London and Washington.

Flexner, J.L., J. B. Fleisher & A. LaViolette
2008 Bead grinders and early household economy: analysis of an assemblage from Pemba Island, Tanzania, 7th-10th centuries AD. *Journal of African Archaeology* Vol. 6 (2): 161-181.

Freeman-Grenville, G. S. P.
1962 *The East African coast: select documents from the first to the earlier nineteenth century.* Clarendon Press, Oxford.

Fritz, G. J. (editor)
1995 *Laboratory Guide to Archaeological Plant Remains from Eastern North America.*

Fryxell, P. A.
1979 *The natural history of the cotton tribe (Malvaceae, Tribe Gossypieae).*Texas A&M University Press, College Station, TX.

Fuller, D. Q
1999 *The emergence of agricultural societies in South India: botanical and archaeological perspectives.* Unpublished PhD thesis, University of Cambridge.
2000 Fifty years of archaeobotanical studies in India: laying a solidfoundation. In *Indian archaeology in retrospect. Volume III: Archaeology and interactive disciplines*, edited by S. Settar and R. Korisettar, pp. 247-363. Publication for the Indian Council for Historical Research, New Delhi.
2003 African crops in prehistoric South Asia: a critical review. In *Foods, Fuel and Fields: Progress in African Archaeobotany*, edited by K. Neumann, S. Kahlheber and A. Butler, pp. 239-271. Heinrich Barth Institut, Köln, Germany.
2004 Early Kushite Agriculture: Archaeobotanical Evidence from Kawa. *Sudan & Nubia Bulletin* 8:70-74.

Fuller, D., N . Boivin, T. Hoogervorst & R. Allaby
2011. Across the Indian Ocean: the prehistoric movement of plants and animals. *Antiquity* 85: 544-58.

Fuller, D., C. Stevens and M. McClatchie
2005 Routine Activities, Tertiary Refuse and Labor Organization: Social Inferences from Everyday Archaeobotany. In *Ancient Plants and People: Contemporary Trends in Archaeobotany*, edited by M. Madella and M. Savard, pp. In Press. University of Arizona Press, Tempe.

Galvin, K. F.
1987 Forms of finance and forms of production: the evolution of specialized livestock production in the ancient Near East. In *Specialization, exchange and complex societies.*, edited by E. Brumfiel and T. Earle, pp. 119-129. Cambridge University Press, Cambridge.

Garlake, P. S.
1966 *The Early Islamic Architecture of the East African Coast.* Oxford University Press, London & Nairobi.
1973 *Great Zimbabwe.* Thames and Hudson, London.
1978 Pastoralism and Zimbabwe. *Journal of African History* 23(4):443-461.

Gilbert, E. and J. T. Reynolds
2004 *Africa in world history: from prehistory to the present.* Pearson/Prentice Hall, Upper Saddle River, NJ.

Goldman, H. V. and M. Walsh
2002 Is the Zanzibar leopard (Panthera pardus adersi) extinct? *the Journal of East African Natural History* 91(1&2):15-25.

Goody, J.
1982 *Cooking, cuisine and class.* Cambridge University Press, Cambridge.

Grab, S., L. Scott, L. Rossouw and S. Meyer
2005 Holocene palaeoenvironment inferred from a sedimentary sequence in the Tsoaing River Basin, Western Lesotho. *Catena* 61- 61(1- 1):49-62- 62.

Grubben, G. J. H. and O. A. Denton (editors)
2004 *Plant Resources of Tropical Africa 2. Vegetables.* PROTA Foundation & Backhuys Publishers, Wageningen, Netherlands.

Gunn, B. F.
2004 The phylogeny of the Cocoeae (Arecaceae) with emphasis on *Cocos nucifera*. *Annals of the Missouri Botanical Garden* 91:505-522.

Gunn, C. R. and J. V. Dennis
1976 *World guide to tropical drift seeds and fruits*. Quadrangle, New York.

Haaland, R.
1994-95 Dakawa: an early Iron-Age Site in the Tanzanian hinterland. In *The Growth of Farming Communities in Africa South of the Equator*, edited by J. E. G. Sutton, pp. 238-247. vol. Azania Special Volume XXIX-XXX.

Hachten, H.
1970 *Best of regional African cooking*. Hippocrene, New York.

Hamdun, S. and N. King
1998 *Ibn Battuta in black africa*. Markus Wiener, Princeton.

Hamilton, A. C.
1982 *Environmental History of East Africa: a study of the Quaternary*. Academic Press, London & New York.

Harlan, J. R.
1993 The Tropical African Cereals. In *The Archaeology of Africa: Food, Metal and Towns*, edited by T. Shaw, P. J. J. Sinclair, B. Andah and A. Okpoko, pp. 53-60. Routledge, London.
1997 Food Crops. In *Encyclopedia of Precolonial Africa: archaeology, history, languages, cultures, and environments*, edited by J. O. Vogel, pp. 225-227. Altamira, Walnut Creek, California.

Harlan, J. R., J. M. J. De Wet and A. B. L. Stemler (editors)
1976 *Origins of African plant domestication*. Mouton, The Hague.

Harries, H. C.
1978 The evolution, dissemination and classification of Cocos nucifera L. *The Botanical Review* 44(3):265-319.
1981 Germination and taxonomy of the coconut. *Annals of Botany* 48:873-883.
1992 Biogeography of the Coconut Cocos nucifera L. *Principes* 36:155-162.

Harvey, E. L. and D. Q Fuller
2005 Investigating crop processing using phytolith analysis: the example of rice and millets. *Journal of Archaeological Science* 32:739-752.

Hassan, F. A.
1997 Holocene Palaeoclimates of Africa. *African Archaeological Review* 14(4):213-230.

Hastorf, C. A.
1993 *Agriculture and the onset of political inequality before the Inka*. Cambridge University Press, Cambridge.

Hather, J. G.
1994 *Tropical archaeobotany: applications and new developments*. One World Archaeology: 22. Routledge, London; New York.
2000 *Archaeological parenchyma*. Archetype publications, London.

Hayden, B.
2001 Fabulous feasts: a prolegomenon to the importance of feasting. In *Feasts: Archaeological and ethnographic perspectives on food, politics, and power*, edited by M. Dietler and B. Hayden, pp. 23-64. Smithsonian Institution Press, Washington.

Heath, D. B.
2000 *Drinking Occasions: Comparative perspectives on alcohol and culture*. Taylor and Francis, Philadelphia.

Helm, R., A. Crowther, C. Shipton, A. Tengeza, D. Fuller, and N. Boivin
2012 Exploring Agriculture, Interaction and Trade on the Eastern African Littoral: Preliminary Results from Kenya. *Azania: Archaeological Research in Africa* 47(1):39–63

Hillman, G. C.
1973 Crop husbandry and food production: modern basis for the interpretation of plant remains. *Anatolian Studies* 23:241-244.
1981 Reconstructing crop husbandry practices from charred remains of plants. In *Farming Practices in British Prehistory*, edited by R. J. Mercer, pp. 123-162. Edinburgh University Press, Edinburgh.
1984 Interpretation of archaeological plant remains: the application of ethnographic evidence from Turkey. In *Plants and Ancient Man: Studies in Palaeoethnobotany*, edited by W. van Zeist and W. A. Casparie, pp. 1-42. proceedings of the Sixth Symposium of the International Work Group for Palaeoethnobotany, Groningen, 30 May-3 June 1983. A.A. Balkema, Rotterdam.

Hirth, K.
2005 The city in Prehispanic central Mexico. In *Early Cities: New perspectives on pre-industrial urbanism*, National Academy of Sciences, Washington D.C.

Holm, L.
1977 *The World's worst weeds: distribution and biology*. Published for the East-West Center by the University Press of Hawaii, Honolulu.

Horton, M.
1996 *Shanga: a Muslim trading community on the East African Coast.* British Institute in Eastern Africa, Nairobi.

Horton, M. and J. Middleton
2000 *The Swahili: the social landscape of a mercantile society.* The peoples of Africa. Blackwell Publishers, Oxford; Malden, Mass.

Horton, M. and N. Mudida
1993 Exploitation of Marine Resources: Evidence for the Origin of Swahili Communities of East Africa. In *The Archaeology of Africa: Food, Metal and Towns*, edited by T. Shaw, P. J. J. Sinclair, B. Andah and A. Okpoko, pp. 673-693. Routledge, London.

Huffman, T. N.
1972 The rise and fall of Zimbabwe. *Journal of African History* 13:353-366.

Huntingford, G. W. B.
1980 *The Periplus of the Erythraean Sea.* Hakluyt Society, London.

Johnson, A. and T. Earle
2000 *The Evolution of Human Societies.* 2nd Edition ed. Stanford University Press, Stanford.

Jones, G. E. M.
1984 Interpretation of archaeological plant remains: ethnographic models from Greece. In *Plants and Ancient Man: Studies in Palaeoethnobotany: proceedings of the Sixth Symposium of the International Work Group for Palaeoethnobotany, Groningen, 30 May-3 June 1983*, edited by W. van Zeist and W. A. Casparie, pp. 43-62. A.A. Balkema, Rotterdam.

Jones, M. K.
1985 Archaeobotany beyond subsistence reconstruction. In *Beyond domestication in prehistoric Europe: investigations in subsistence archaeology and social complexity*, edited by G. Barker and C. Gamble, pp. 107-128. Academic Press, London.

Jonsson, J.
1998 *Early plant economy in Zimbabwe.* Studies in African Archaeology:16. Department of Archaeology and Ancient History, Uppsala University, Uppsala, Sweden.

Kabuye, C.
1991 The Importance of Multidisciplinary Work on the Transition and Development of Agriculture in East Africa. In *Origins and development of agriculture in East Africa: the ethnosystems approack to the study of early food production in Kenya*, edited by R. E. Leakey and L. J. Slikkerveer, pp. 215-217. Studies in Technology and Social Change No. 19. Iowa State University, Ames.

Kelly, L. S.
2001 A case of ritual feasting at the Cahokia Site. In *Feasts: Archaeological and ethnographic perspectives on food, politics, and power.*, edited by M. Dietler and B. Hayden, pp. 334-367. Smithsonian Institution Press, Washington.

Kenoyer, J. M.
2005 The origin and character of Indus urbanism: new perspectives and challenges. In *Early Cities: new perspectives on pre-industrial urbanism*, National Academy of Sciences - Washington, D.C.

Kirch, P. V.
1994 *The wet and the dry: irrigation and agricultural intensification in Polynesia.* University of Chicago Press, Chicago.

Kirkman, J. S.
1954 *The Arab city of Gedi: excavations at the Great Mosque, architecture and finds.* Oxford University Press, London.
1966 *Men and Monuments on the East African Coast.* Frederick A. Praeger, New York.

Klapwijk, M.
1974 A preliminary report on pottery from the North-eastern Transvaal, South Africa. *South African Archaeological Bulletin* 29:19-23.

Koenders, L.
1992 *Fauna of Pemba Island: a checklist of animals occurring on Pemba Island, Tanzania.* Wildlife Conservation Society of Tanzania Publication Number 1.
1992 *Flora of Pemba Island: a checklist of species.* Wildlife Conservation Society of Tanzania Publication 2.

Kramer, C.
1994 Scale, Organization, and Function in Village and Town. In *Archaeological views from the countryside: village communities in early complex societies*, edited by G. M. Schwarz and S. E. Falconer. Smithsonian Institution Press, Washington & London.

Kuklick, H.
1991 Contested Monuments: The politics of archaeology in Southern Africa. In *Colonial Situations: Essays on the Contextualization of Ethnographic Knowledge.*, edited by G. W. J. Stocking, pp. 135-169. University of Wisconsin Press, Madison.

Kusimba, C. M.
1997 Swahili and the coastal city-states. In *Encyclopedia of Precolonial Africa*, edited by J. O.

Vogel, pp. 507-512. Altamira Press, Walnut Creek, CA.
1999 *The rise and fall of Swahili states*. AltaMira Press, Walnut Creek, Calif.

Kusimba, C. M. and S. B. Kusimba
2000 Hinterlands and cities: archaeological investigations of economy and trade in Tsavo, southeastern Kenya. *Nyame Akuma* 54:13-24.

Kyule, D. M.
1991 Plant remains from a Sirikwa culture site at Hyrax Hill, Nakuru, Kenya. *Nyame Akuma* 36:8-10.

LaViolette, A.
2000 Swahili archaeology on Pemba Island, Tanzania: Pujini, Bandari ya Faraji, and Chwaka 1997-1998. *Nyame Akuma* 53:50-61.

LaViolette, A., W. B. J. Fawcett, N. J. Karoma, P. R. Schmidt and
1989 The coast and the hinterland: University of Dar es Salaam archaeological field schools, 1987-88. *Azania* XXXII:38-46.

LaViolette, A. and J. Fleisher
1995 Reconnaissance of sites bearing triangular incised (Tana tradition) ware on Pemba Island, Tanzania. *Nyame Akuma* 56:36-43.
2005 The archaeology of Sub-Saharan urbanism: cities and their countrysides. In *African archaeology: a critical introduction*, edited by A. B. Stahl, pp. 327-352. Blackwell, Oxford.2009 The Urban History of a Rural Place: Swahili Archaeology on Pemba Island, Tanzania, 700-1500 AD. *The International Journal of African Historical Studies* 42 (3): 433-455

LaViolette, A., J. Fleisher and B. B. B. Mapunda
2003 *Pemba Archaeological Project Preliminary Report: First Season, June-August 2002.* Department of Archives, Museums and Antiquities, Ministry of Education and Culture.
2004 *Pemba Archaeological Project Preliminary Report: Second Season, June-July 2004.* Department of Archives, Museums and Antiquities, Ministry of Education and Culture.

Lewis, W. H. and M. P. F. Elvin-Lewis
1977 *Medical Botany: plants affecting man's health.* John Wiley & Sons, New York.

Liesner, R.
2005 Field Techniques Used by Missouri Botanical Garden. vol. 2005.

Liu, R. K.
1978 Spindle whorls, Part I: some comments and speculations. *The Bead Journal* 3(3):87-103.

Lloyd, M., J. K. Zar and J. B. Kar
1968 On the calculation of information-theoretical methods of diversity. *American Midlands Naturalist* 79:257-272.

Lyman, R. L.
1987 Archaeofaunas and butchery studies: A taphonomic perspective. In *Advances in Archaeological Method and Theory (Vol. 10)*, edited by M. B. Schiffer, pp. 249–337. Academic Press, New York.

MacDonald, K.
1995 Analysis of the mammalian, avian and reptilian remains. In *Excavations at Jenné-jeno, Hambarketolo, and Kaniana (Inland Niger Delta, Mali), the 1981 season*, edited by S. K. McIntosh, pp. 291-318. University of California Press, Berkeley, Los Angeles.

MacLean, R. and T. Insoll
2003 Archaeology, luxury, and the exotic: examples from Islamic Gao (Mali) and Bahrain. In *Luxury Foods*, edited by M. van der Veen, pp. 558-570. *World Archaeology* 34(3). Abingdon, Taylor and Francis.

Maggs, T. O. C. and V. Ward
1984 Early iron age sites in the Muden area of Natal. *Annals of the Natal Museum* 26:105-140.

Mapunda, B. B. B.
2002 Iron metallurgy along the Tanzanian Coast. In *Southern Africa and the Swahili World.*, edited by F. Chami and G. Pwiti, pp. 76-88. Dar es Salaam University Press, Dar es Salaam.

Marcus, J. and J. Sabloff
2005 Organizers. In *Early Cities: New perspectives on pre-industrial urbanism*, National Academy of Sciences, Washington, D.C.

Marshall, F.
1998 Early food production in Africa. *The Review of Archaeology* 19(2):47-58.

Marshall, F. and E. A. Hildebrand
2002 Cattle before crops: The beginnings of food production in Africa16(2):99-143.

Martin, A. C. and W. D. Barkley
1961 *Seed identification manual.* University of California Press, Berkeley.

Martin, G. J.
1995 *Ethnobotany: a methods manual.* 1st ed. "People and plants" conservation manuals; v. 1. Chapman & Hall, London; New York.

Maxon, R. M.
1994 *East Africa: an introductory History*. 2nd Revised ed. West Virginia University Press, Morgantown.

McCann, J.
2001 Maize and Grace: History, Corn, and Africa's New Landscapes, 1500-1999. *Comparative Studies in Society and History* 43(2):246-272.

McIntosh, R. and S. K. McIntosh
2003 Early urban configuration in the Inland Niger Delta: Clustered Cities and landscapes of power. In *The social construction of ancient cities*, edited by M. L. Smith, pp. 103-120. Smithsonian Institution Press, Washington, D.C.

McIntosh, R. J.
2000 Clustered cities of the middle Niger: Alternative routes to authority in prehistory. In *Africa's Urban Past*, edited by D. M. Anderson and R. Rathbone, pp. 19-35. James Currey, Oxford.

McIntosh, S. K. (editor)
1995a *Excavations at Jenné-Jeno, Hambarketolo, and Kaniana (Inland Niger Delta, Mali), the 1981 season.* University of California Press, Berkeley and Los Angeles.
1995b Paleobotanical and human osteological remains. In *Excavations at Jenné-Jeno, Hambarketolo, and Kaniana (Inland Niger Delta, Mali), the 1981 season*, edited by S. K. McIntosh, pp. 348-359. University of California Press, Berkeley and Los Angeles.
1999 *Beyond chiefdoms: pathways to complexity in Africa*. New directions in archaeology. Cambridge University Press, Cambridge, U.K.; New York.

McIntosh, S. K. and R. J. McIntosh
1980 Jenne-jeno, an ancient African City. *Archaeology* 33:8-14.
1984 The early city in west Africa: towards an understanding. *African Archaeological Review* 2:73-98.
1993 Cities without citadels: understanding urban origins along the middle Niger. In *The Archaeology of Africa: Foods, Metals and Towns*, edited by T. S. e. al., pp. 622-641. Routledge, London.

McParland, H. and S. Walshaw
2015 Swahili Plant Use and Activity Areas at Songo Mnara, Tanzania: A Case Study at Macro and Micro Scales. In *Società dei Naturalisti e Matematici di Modena*, Suppl. Vol CXLVI Proceedings of the IWAA8 pp. 193-195.

Middleton, J.
1992 *The world of the Swahili: an African mercantile civilization*. Yale University Press, New Haven.

Miksicek, C. M. I., pp., edited by Michael B. Schiffer. Academic Press.
1987 Formation Processes of the Archaeobotanical Record. In *Advances in Archaeological Method and Theory, Vol. 10*, edited by M. B. Schiffer, pp. 211-247. Academic Press, New York.

Miller, N. F.
1988 Ratios in paleoethnobotanical analysis. In *Current Paleoethnobotany: Analytical methods and cultural interpretations of archaeological plant remains*, edited by C. A. Hastorf and V. S. Popper, pp. 72-85. University of Chicago Press, Chicago.

Miller, N. F. and T. L. Smart
1984 Intentional burning of dung as fuel: a mechanism for the incorporation of charred seeds into the archaeological record. *Journal of Ethnobiology* 4:15-28.

Montgomery, F. H.
1977 *Seeds and fruits of plants of eastern Canada and northeastern United States*. University of Toronto Press, Toronto; Buffalo.

Morcote-Rios, G. and R. Bernal
2001 Remains of palms (Palmae) at archaeological sites in the New World: A review *The Botanical Review* 67(3):309-350.

Mudida, N.
1996 Subsistence at Shanga: the faunal record. In *Shanga: the archaeology of a Muslim trading community on the coast of East Africa*, edited by M. Horton, pp. 378-393. vol. BIEA Memoir 14. British Institute in East Africa, London.

Mutoro, H. W.
1998 Precolonial trading systems of the East African interior. In *Transformations in Africa: essays on Africa's later past*, edited by G. Connah, pp. 186-203. Leicester University Press, London & Washington.

Muzuka, A. N. N., M. Ryner and K. Holmgren
2004 12,000 year preliminary results of the stable nitrogen and carbon isotope record from the Empakai Crater lake sediments, northern Tanzania. *Journal of African Earth Sciences* 40(5):293-303.

Mwale, D. and T. Y. Gar
2005 Wavelet analysis of variability, teleconnectivity, and predictability of the

September-November East African Rainfall. *Journal of Applied Meteorology* 44(2):256-269.

Nesbitt, R. M. A.
1997 *Archaeobotanical identification of Near Eastern grass caryopses.* Unpublished Ph.D. Thesis, University College London.

Netting, R. M.
1993 *Smallholders, householders: Farm families and the ecology of intensive, sustainable agriculture.* Stanford University Press, Stanford CA.

Neumann, K.
2003 The late emergence of agriculture in Sub-Saharan Africa: Archaeobotanical evidence and ecological considerations. In *Food, Fuel and Fields: Progress in African Archaeobotany*, edited by K. Neumann, A. Butler and S. Kahlheber, pp. 71-92. Heinrich-Barth Institut, Köln.

2005 The romance of farming: plant cultivation and domestication in Africa. In *African archaeology: a critical introduction*, edited by A. B. Stahl, pp. 249-275. Blackwell, Malden, MA.

NIAB
2004 *Seed Identification Handbook: Agriculture, horticulture & weeds.* 2nd ed. NIAB, Cambridge.

Northam, R. M.
1975 *Urban Geography.* John Wiley & Sons, New York.

NRC
1996 *Lost Crops of Africa Volume 1: Grains.* National Academy Press, Washington.

Nurse, D. and T. T. Spear
1985 *The Swahili: reconstructing the history and language of an African society, 800-1500.* University of Pennsylvania Press, Philadelphia.

Ohler, J. G.
1999a The coconut palm. In *Modern coconut management: palm cultivation and products*, edited by J. G. Ohler, pp. 11-32. Food and Agriculture Organization, Leiden.
1999b Historical background. In *Modern coconut management: palm cultivation and products*, edited by J. G. Ohler, pp. 1-10. FAO, Leiden.

Ominde, M.
1975 *African cookery book.* Kenway, Naironi.

Palmer, C. and M. van der Veen
2002 Archaeobotany and the social context of food. *Acta Palaeobotanica* 42(2):195- 202.

Pauketat, T. R., L. S. Kelly, G. J. Fritz, N. H. Lopinot, S. Elias and E. Hargrave
2002 The Residues of Feasting and Public Ritual at Early Cahokia. *American Antiquity* 67(2).

Pawlowicz, M.
2011 *Finding Their Place in the Swahili World: An archaeological exploration of southern Tanzania.* Unpublished PhD thesis, Department of Anthropology, University of Virginia

Pawlowicz, M., J. Stoetzel and Macko, S.
2014 Environmental Archaeology at Mikindani, Tanzania: Towards a Historical Ecology of the Southern Swahili Coast. *Journal of African Archaeology*, 12(2): 119-139

Pearce, M. F. B.
1967 [1920] *Zanzibar: The island metropolis of Eastern Africa.* Frank Cass&Co., London.

Pearsall, D. M.
2000 *Paleoethnobotany: a handbook of procedures.* 2nd ed. Academic Press, San Diego.

Phillipson, D. W., A. Reynolds, S. Boardman, N. Finneran and J. Philips
1996 B.I.E.A. excavations at Aksum, Northern Ethiopia, 1995. *Azania* 31:99-147.

Pikirayi, I.
2001 *The Zimbabwe Culture: Origins and decline of southern Zambezian states.* Altamira, Walnut Creek, CA.

Popper, V. S.
1988 Selecting quantitative measurements in paleoethnobotany. In *Current Paleoethnobotany: Analytical methods and cultural interpretations of archaeological plant remains*, edited by C. A. Hastorf and V. S. Popper, pp. 53-71. University of Chicago Press, Chicago.

Portères, R.
1976 African cereals: Eleusine, Fonio, Black Fonio, Teff, Brachiaria paspalum, Pennisetum, and African rice. In *Origins of African Plant Domestication*, edited by J. R. Harlan, J. M. J. De Wet and A. B. L. Stemler, pp. 409-452. Mouton, The Hague.

Posnansky, M.
1984 The ethnoarchaeology of farm shelters at Hani, Ghana. *Anthroquest* 30(11-12).

Pritchard, J. M.
1975 *A Geography of East Africa.* Evans Brothers, London.

Pwiti, G.
1996 Settlement and subsistence in prehistoric farming communities in the mid-Zambezi Valley, northern Zimbabwe. *South African Archaeological Bulletin* 51:3-6.

Reddy, S. N.
1997 If the threshing floor could talk: integration of agriculture and pastoralism during the late Harappan in Gujarat, India. *Journal of Anthropological Archaeology* 16:162-187.

Rehm, S. and G. Espig
1991 *The Cultivated Plants of the Tropics and Subtropics: Cultivation, economic value, utilization.* Verlag Josef Margraf, Netherlands.

Reid, A. and R. Young
2000 Pottery abrasion and the preparation of African grains. *Antiquity* 74(283):101-112.

Robertshaw, P.
1990 A history of African archaeology: an introduction. In *A History of African Archaeology.*, edited by P. Robertshaw, pp. 3-12. Currey, London.

Robertshaw, P. and W. Wetterstrom
1989 Plant Remains from Gogo Falls. *Nyame Akuma* 31:25-27.

Rouguelle, A.
1996 Medieval trade networks in the western Indian Ocean (8-14th centuries): Some reflections from the distribution pattern of Chinese imports in the Islamic World. In *Tradition and Archaeology: Early Maritime Contacts in the Indian Ocean*, edited by H. P. Ray and J.-F. Salles, pp.159-180. Manohar Publishers and Distributors, New Delhi.

Rowley-Conwy, P., W. J. Deakin and C. H. Shaw
1997 Ancient DNA from archaeological sorghum (Sorghum bicolor) from Qasr Ibrim, Nubia: implications for domestication and evolution and a review of the archaeological evidence. *Sahara: prehistory and history of the Sahara* 9:23-34.

Russell, J. M. and T. C. Johnson
2005 A high resolution geochemical record for Lake Edward, Uganda Congo, and the timing and causes of African draught during the late Holocene. *Quaternary Science Review* 24:1375-1389.

Sauer, J. D.
1967 *Plants and Man on the Seychelles Coast: A study in historical biogeography*. University of Wisconson Press, Madison, WI.

Schlüter, T.
1997 *Geology of East Africa*. Gebrüder Borntraeger, Berlin, Stuttgart.

Schwarz, G. M. and S. E. Falconer (editors)
1994 *Archaeological Views from the countryside: village communities in early complex societies.* Smithsonian Institution Press, Washington and London.

Shipton, C., R. Helm, N. Boivin, A. Crowther, P. Austin, and D. Fuller
2013 Intersections, Networks and the Genesis of Social Complex ity on the Nyali Coast of East Africa. African Archaeological Review 30(4):427–453.

Simmonds, N.W. and K. Shepherd
1952 An Asian Banana (Musa acuminata) in Pemba, Zanzibar Protectorate. *Nature* 169 (22 March): 507 - 508.

Sinclair, P. J. J.
1987 *Space, time and social formation: a territorial approach to the archaeology and anthropology of Zimbabwe and Mozambique c. 0-1700 AD.* Societas Archaeologica Upsaliensis, Uppsala.

Sinclair, P. J. J. and N. T. Håkansson
2000 The Swahili City-State Culture. In *A Comparative Study of Thirty City-State Cultures*, edited by M. H. Hansen, pp. 463-482. The Royal Danish Academy of Sciences and Letters, Copenhagen.

Sjoberg, G.
1960 *The Preindustrial City, Past and Present.* Free Press, Glencoe.

Smith, M. L.
2003 Introduction: The Social Construct of Ancient Cities. In *The Social Construction of Ancient Cities*, edited by M. L. Smith, pp. 1-36. Smithsonian, Washington.

Smith, W.
2001 When method meets theory: the use and misuse of cereal producer/consumer models in archaebotany. In *Environmental Archaeology: Meaning and Purpose*, edited by U. Albarella, pp. 283-298. Klewer Academic, Dordrecht.

Soper, R.
1997 Eastern African Terraced-Irrigation Systems. In *Encyclopedia of Precolonial Africa: archaeology, history, languages, cultures, and environments*, edited by J. O. Vogel, pp. 227-231. Walnut Creek, California, Altamira.

Stager, J. C., D. Ryves, B. F. Cumming, L. D. Meeker and J. Beer
2005 Solar variability and the levels of Lake Victoria, East Africa, during the last millenium. *Journal of Paleolimnology* 33(2):243-251.

Stahl, A. B.
1984 History and critique of investigations into early African agriculture. In *From hunters to farmers*, edited by J. D. Clark and S. A. Brandt, pp. 9-21. University of California Press, Berkeley.

Stevens, C. J.
2003 An investigation of agricultural consumption and production models for prehistoric and Roman Britain. *Environmental Archaeology* 8(1):61-76.

Stevens, P. F.
2001 onwards Angiosperm Phylogeny Website. Version 5.

Stiles, D.
1992 The ports of east Africa, the Comoros and Madagascar: their place in Indian Ocean trade from 1-1500 AD. *Kenya Past and Present* 24:1-10.

Stoetzel, J.
2014. Towards a Swahili Historical Ecology: Phytolith-based Analysis in Coastal Eastern Africa since AD 600. Unpublished PhD thesis, Department of Anthropology, University of Virginia.

Street-Perrott, A. F. and A. R. Perrott
1990 Abrupt climate fluctuations in the tropics: the influence of Atlantic Ocean circulation. *Nature* 343:607–612.

Sulas, F. and M. Madella
2012. Archaeology at the micro-scale: micromorphology and phytoliths at a Swahili stonetown. *Archaeological and Anthropological Sciences* 4(2):145-159.

Summers, R.
1958 *Inyanga: Prehistoric settlements in Southern Rhodesia.* Cambridge University Press, Cambridge.

Sutton, J. E. G.
1978 Engaruka and its waters. *Azania* 13:37-70.
1986 The irrigation and manuring of the Engaruka field system. *Azania* 21:27-51.
1987 Hyrax hill and the Sirikwa. *Azania* 22:1-36.

Trigger, B.
1972 Determinants of urban growth in pre-industrial societies. In *Man, Settlement, and Urbanism*, edited by P. J. Ucko, R. Tringham and G. W. Dimbleby, pp. 575-595. Duckworth, London.

Ucko, P. J., R. Tringham and G. W. Dimbleby (editors)
1972 *Man, settlement and urbanism.* Duckworth, London.

van der Veen, M.
1991 Consumption or production: agriculture in the Cambridgeshire Fens. In *New light on early farming*, edited by J. M. Renfrew, pp. 349-361. vol. Proceedings of the 7th symposium of the international work group for paleoethnobotany. Edinburgh University Press, Edinburgh.
2003a Trade and diet at Roman and medieval Quseir al-Qadim, Egypt. A preliminary report. In *Food, Fuel and Fields: Progress in African Archaeobotany*, edited by K. Neumann, A. Butler and S. Kahlheber, pp. 207-212. Heinrich Barth Instutut, Koln.
2003b *World Archaeology: Luxury Foods. Volume 34 (3).* Routledge, Milton Park.

van Neer, W.
1995 Analysis of the fish remains. In *Excavations at Jenné-jeno, Hambarketolo, and Kaniana (Inland Niger Delta, Mali), the 1981 season*, edited by S. K. McIntosh, pp. 319-347. University of California Press, Berkeley, Los Angeles.

Vollesen, K.
1987 The native species of Gossypium (Malvaceae) in Africa, Arabia, and Pakistan. *Kew Bulletin* 42(2):337-349.

Wagner, G. E.
1988 Comparability among recovery techniques. In *Current Paleoethnobotany: Analytical methods and cultural interpretations of archaeological plant remains*, edited by C. A. Hastorf and V. S. Popper, pp. 17-35. University of Chicago Press, Chicago.

Walsh, M.
1995 Eating bats on Pemba Island: a local innovation or cultural borrowing? *Mvita: newsletter of the regional centre for the study of archaeology in eastern and southern Africa* 6:15-18.
2009 The use of wild and cultivated plants as famine foods on Pemba Island, Zanzibar. *Études Océan indien*, 42-43:217-241.

Walshaw, S.
2005 Finding *mtama*: archaeobotanical preservation and recovery in tropical eastern Africa. Paper presented at the Chacmool Archaeological Conference, University of Calgary, Calgary, Alberta, Canada.
2010 Converting to Rice: Urbanization, Islamization and Crops on Pemba Island, Tanzania AD 700-1500. *World Archaeology* 42(1): 137-153.
2015. Cooking and Crafting at Songo Mnara, Tanzania (14th-15th C): Evidence from

Macrobotanicals. In *Società dei Naturalisti e Matematici di Modena*, Suppl. Vol CXLVI Proceedings of the IWAA8 pp. 121-123.

Ward, R. G. and M. Brookfield
1992 The Dispersal Of The Coconut - Did It Float Or Was It Carried To Panama19(5):467-480.

Watson, A. M.
1983 *Agricultural Innovation in the early Islamic World: The diffusion of crops and farming techniques, 700-1100*. Cambridge University Press, Cambridge.

Wattenmaker, P.
1998 *Household and State in Upper Mesopotamia: specialized economy and the social uses of goods in an early complex society*. Smithsonian Institution Press, Washington.

Weber, M.
1962 *The City*. Translated by D. Martindalen and G. Neuwirth. Collier Books, New York.

Wendrich, W. Z., R. S. Tomber, S. E. Sidebotham, J. A. Harrell, R. T. J. Cappers and R. S. Bagnall
2003 Berenike Crossroads: the integration of information. *JESHO* 46(1):46-87.

Wetterstrom, W.
1991 Fleshing out early agriculture with ancient seeds. In *Origins and development of agriculture in East Africa: the ethnosystems approach to the study of early food production in Kenya*, edited by R. E. Leakey and L. J. Slikkerveer, pp. 219-234. Studies in Technology and Social Change No. 19. Iowa State University, Ames.

Wetterstrom, W. and H.T. Wright
2007. Appendix C: A contribution to the paleoethnobotany of the central highlands of Madagascar. In *Early State Formation in Central Madagascar: An Archaeological Survey of Western Avaradrano* (ed. H.T. Wright). Ann Arbor: Museum Memoirs Number 43, pp. 281-288.

White, F.
1983 *The vegetation of Africa*. UNESCO, Paris.

Wiessner, P. and W. Schiefenhövel (editors)
1996 *Food and the status quest: an interdisciplinary perspective*. Berghahn Books, Oxford.

Wigboldus, J. S.
1994-95 The Spread of Crops in Sub-Equatorial Africa during the Early Iron Age: A 'Minimalist' View Based Primarily on Documentary Evidence from the Indian Ocean Side. *Azania* XXIX-XXX:121-129.

Williams, R. O.
1949 *The useful and ornamental plants of Zanzibar and Pemba*. St Ann's Press, Timperly, Altrincham.

Wilson, T. H.
1982 Spatial analysis and settlement patterns on the East African coast. *Paideuma* 28:201-219.

Wilson, T. H. and A. L. Omar
1997 Archaeological Investigations at Pate. *Azania* XXXII:31-76.

Wittfogel, K. A.
1957 *Oriental Despotism: A comparative study of total power*. Yale University Press, New Haven.

Wright, H. T.
1984 Early seafarers of the Comoro Islands: the Dembeni phase of the IXth-Xth centuries AD. *Azania* XIX:13-59.
1992 Early Islam, oceanic trade and town development on Nzwani: The Comorian archipelago in the XIth-XVth Centuries AD. *Azania* XXVII:111-114.
1993 Trade and politics on the eastern littoral of Africa, AD 800-1300. In *The Archaeology of Africa: Food, metals and towns*, edited by T. Shaw, P. J. J. Sinclair, B. Andah and A. Okpoko, pp. 658-672. vol. London. Routledge.

Wylie, A.
1985 The reaction against analogy. In *Advances in Archaeological Method and Theory*, edited by M. B. Schiffer, pp. 63-111. vol. 8. Academic Press, New York.

Young, R. and G. Thompson
1999 Missing plant foods? Where is the archaeobotanical evidence for sorghum and finger millet in East Africa? In *The exploitation of plant resources in ancient Africa*, edited by M. van der Veen, pp. 63-72. Klewer Plenum, New York.

Zeder, M. A.
1991 *Feeding Cities: Specialized animal economy in the ancient near east*. Smithsonian Institution Press, Washington & London.
1999 Understanding urban process through the study of specialized subsistence economy in the Near East. *Journal of Anthropological Archaeology* 7(1):1-55.
2003 Food Provisioning in Urban Societies: A view from northern Mesopotamia. In *The Social Construction of Ancient Cities*, edited by M. L. Smith, pp. 156-183. Smithsonian Institution Press, Washington.

Zhao, Z.
1998 The middle Yangtze region in China is one place where rice was domesticated: phytolith evidence from the Diaotonghuan Cave, Northern Jiangxi. *Antiquity* 72:885-897.

Zohary, D. and M. Hopf
2000 *Domestication of plants in the old world.* Third ed. Oxford University Press, Oxford.

www.ingramcontent.com/pod-product-compliance
Lightning Source LLC
LaVergne TN
LVHW071219210926
848482LV00015B/853